A Series In

The History of Modern Physics, 1800-1950

The History of Modern Physics, 1800-1950

TITLES IN SERIES

INTRODUCTORY NOTE

The Tomash series in the History of Modern Physics offers the opportunity to follow the evolution of physics from its classical period in the nineteenth century when it emerged as a distinct discipline, through the early decades of the twentieth century when its modern roots were established, into the middle years of this century when physicists continued to develop extraordinary theories and techniques. The one hundred and fifty years covered by the series, 1800 to 1950, were crucial to all mankind not only because profound evolutionary advances occurred but also because some of these led to such applications as the release of nuclear energy. Our primary intent has been to choose a collection of historically important literature which would make this most significant period readily accessible.

We believe that the history of physics is more than just the narrative of the development of theoretical concepts and experimental results: it is also about the physicists individually and as a group—how they pursued their separate tasks, their means of support and avenues of communication, and how they interacted with other elements of their contemporary society. To express these interwoven themes we have identified and selected four types of works: reprints of "classics" no longer readily available; original monographs and works of primary scholarship, some previously only privately circulated, which warrant wider distribution; anthologies of important articles here collected in one place; and dissertations, recently written, revised, and enhanced. Each book is prefaced by an introductory essay written by an acknowledged scholar, which, by placing the material in its historical context, makes the volume more valuable as a reference work.

The books in the series are all noteworthy additions to the literature of the history of physics. They have been selected for their merit, distinction, and uniqueness. We believe that they will be of interest not only to the advanced scholar in the history of physics, but to a much broader, less specialized group of readers who may wish to understand a science that has become a central force in society and an integral part of our twentieth-century culture. Taken in its entirety, the series will bring to the reader a comprehensive picture of this major discipline not readily achieved in any one work. Taken individually, the works selected will surely be enjoyed and valued in themselves.

A Series In

The History of Modern Physics, 1800-1950

VOLUME I

ALSOS

ALSOS

BY

Samuel A. Goudsmit

WITH A

NEW INTRODUCTION BY

R.V. Jones

*Emeritus Professor of Natural
Philosophy, Aberdeen University*

Tomash Publishers

American Institute of Physics
A̲I̲P

ACKNOWLEDGMENTS

The publisher would like to thank Mrs. Samuel A. Goudsmit and Colonel Boris T. Pash for their assistance in preparing this new edition. Numerous photographs originally made available by Brookhaven National Laboratory were furnished by Mrs. Goudsmit; Colonel Pash helped us to identify many of the individuals who were members of the Alsos Mission. We are also indebted to the directors and staff members of the Center for the History of Physics and the Niels Bohr Library at the American Institute of Physics who allowed us to review archival materials on Samuel Goudsmit and also provided us with photographs.

THIRD PRINTING, 1988

This printing is part of a co-publishing agreement between Tomash Publishers and the American Institute of Physics.

Library of Congress Cataloging in Publication Data

Goudsmit, Samuel Abraham, 1902–1978
   Alsos.

   (History of modern physics, 1800–1950; v. 1)
   Originally published: New York: H. Schuman, © 1947.
With new introd.
   Includes index.
   1. World War, 1939–1945—Science. 2. United States. War Dept. Alsos Mission. 3. Atomic bomb—Germany. 4. Nuclear research—Germany. 5. Germany—History—1933–1945. I. Title. II. Series.
D810.S2G6   1983   940.54'8   82–50749
ISBN 0-938228-09-9

Introduction

WHEN in December 1938 Otto Hahn and Fritz Strassmann discovered that the bombardment of uranium nuclei by neutrons appeared to give rise to barium nuclei, they provided the vital clue that within seven years would lead to the uranium fission bomb. Hahn's collaborator, Lise Meitner, and her nephew, Otto Frisch, quickly saw that the explanation of Hahn's results was the fission of the uranium nucleus, and physicists immediately began to wonder whether the fission process released further neutrons which could, in turn, cause other uranium nuclei to split, and whether there would be enough neutrons to give an ever-expanding chain reaction. If so, a bomb of enormous power might be possible. In Germany, England, France, and the United States, physicists alerted their governments; and it seemed to some of us in Britain that at least one physicist in Germany, Siegfried Flügge, by publishing a detailed discussion in *Naturwissenschaften* (Vol. 27, June 9, 1939, pp. 402–410), was trying to warn his colleagues abroad that his own country was seriously considering the exploitation of nuclear energy based on uranium. So, when war broke out in September 1939, it was vital for England and France to watch for German progress. By the time America entered the war two years later, work in England had shown that, given enough effort, a bomb could almost certainly be made.

I had therefore set the monitoring of German activities in nuclear energy as one of my targets when I was appointed to British Intelligence in September 1939. The first direct evidence that the Germans might be experimenting on a fairly large scale came in 1941 from the Norwegian Resistance, who reported that the Germans were ordering large quantities of heavy water; we presumed that this was for the moderator in a fission pile. When we asked for further information, the Norwegians signalled that they would provide it if we could assure them that our interest was not inspired by Imperial Chemical Industries because, their reply ran, "Remember, blood is thicker even than heavy water!"

The further that work on a bomb progressed in the United States, the more reasonable the assumption that there had been corresponding progress in Germany. And with rumors of new German secret weapons rampant, American concern crystallized, in the autumn of 1943, in the form of the Alsos Mission, which was given the duty to "follow immediately in the wake of our armies in the invasion of Europe, for the purpose of determining precisely how much the Germans knew about the atomic bomb and how far they had progressed in its construction" (Foreword).

The Mission was America's first serious effort in Scientific Intelligence. The original Mission (Alsos I) went to Italy in December 1943, but gained little of interest, due in part to the slow progress of the Allied advance toward Rome. The second Mission (Alsos II) was to follow the Allied landings into France and the Low Countries and ultimately into Germany itself. Its scientific head was Dr. Samuel Goudsmit, and this book is his account, written within three years of the end of the war. The statement of the Mission's objective in the preceding paragraph is what he understood his duties to be, although the original terms of reference covered "all principal scientific military developments" (L. R. Groves, *Now It Can Be Told*, 1962; reprint New York: Da Capo, 1975, p. 190). To some extent, both General Groves and Dr. Goudsmit suggest that these more general terms were written to camouflage the Mission's true objective; but since it carried, and sometimes exploited, a Presidential warrant for the much wider field, the Mission's activities were not always without embarrassment to those of us who were already engaged in Scientific Intelligence.

When I first met Goudsmit early in 1944, I gained the impression that he had little idea of what we in Britain had already done by way of nuclear intelligence, and this impression is confirmed in his book where, for example, he says, "Ordinary Intelligence information yielded nothing of value" (p. 10). The facts were that we knew about, and had successfully attacked, the Norwegian heavy water production. Kurt Diebner, one of the senior German physicists assessed the effect of this attack after the war: "It was the elimination of German heavy-water production in Norway that was the main factor in our failure to achieve a self-sustaining atomic reactor before the war ended." (D. Irving, *The Virus House*, London: Kimber, 1967, p. 191). We also had been able to follow the movements of the man who was to be Goudsmit's prime target, Werner Heisenberg, and to establish that he had not been associated with any large scale construction project such as the production of an atomic bomb. Such a production could have required at least three years for its realization. Our knowledge was, incidentally, largely due to Paul Rosbaud, of whom Goudsmit writes so warmly and rightly on pages 185 and 186, and whom we had managed to contact through our Intelligence sources in Switzerland. We had also contrived to approach Niels Bohr in Denmark, and offer him hospitality if ever he chose to join us. An account of how we did so can be found in a memoir by Aage Bohr in *Niels Bohr*, edited by S. Rozental (Amsterdam: North Holland, 1967).

As a result of these contacts, we were fairly confident that the German work was very substantially behind that in America; and we never seriously considered, for example, that the V1 or V2 warheads would be nuclear, even though the V2, in particular, would have been an ideal means of nuclear delivery, and even though we had traced one nuclear physicist, Pascal Jordan, to Peenemünde. So if, as Goudsmit relates on page 13, there was genuine concern on this account in America, we did not share that concern in Britain even though, or perhaps because, we were much nearer to Germany.

I hope that the foregoing comments do not sound unduly carping: they have had to be made in the interests of accuracy and in fairness to the important work done by "ordinary Intelligence" in the years before Goudsmit became involved. They do not detract from the main thrusts of the book which deal with what happened in the Alsos Mission itself and the author's diagnosis of where and why German science was not as effective as science in America and Britain.

Goudsmit's assessment of the state of German nuclear

research and development as revealed by the Alsos Mission is naturally authoritative. It has since been supplemented by David Irving's *The Virus House*, which records that R. Döpel and Heisenberg achieved positive neutron production in a pile designated L-IV because it was the fourth pile built at Leipzig; it used more than half a ton of uranium and one hundred and forty kilograms of heavy water. This was in May 1942, some months ahead of Enrico Fermi in Chicago, but his pile was self-sustaining whereas theirs indicated only that a self-sustaining pile could be made.

The question has often arisen of why the German physicists, having gotten so far, did not proceed seriously to develop the bomb. Heisenberg himself, in an article published in *Nature* (Vol. 160, 1947, pp. 211-215), gave two reasons: (1) the project could not have succeeded because in 1942 the military situation favored only short-term developments and (2) it would have been beyond Germany's scientific and technical resources, anyway, within the duration of the war. Faced with this situation, the German physicists "did not attempt to advocate with the supreme command a great industrial effort for the production of atomic bombs."

Heisenberg's statement is supported by Albert Speer in *Inside the Third Reich* (London: Sphere Books, 1975, pp. 317-320), who stated that the nuclear physicists had told him, in the autumn of 1942, that the project to build a bomb should be scuttled because it could not be achieved within three or four years. Speer authorized work to continue on a uranium engine and thenceforward diverted much of the uranium production to replacing tungsten in heavy-core ammunition. As for what the Nazis would have done if a nuclear bomb had been developed in Germany, Speer said, "I am sure that Hitler would not have hesitated for a moment to employ atomic bombs against England." But, Speer also said, the Nobel Prize winner Philipp Lenard had inspired Hitler to regard relativity and nuclear physics as "Jewish physics," and therefore the prospects for large scale support for these subjects under the Nazis were weakened.

Goudsmit himself attributed the German failure to four

factors (pp. 242–243): (1) complacency among the German physicists, who believed that they had a lead over their Allied counterparts; (2) a deterioration of the Nazi interest in pure science; (3) regimentation in the administrative control of science; and (4) too much respect among their colleagues for "great physicists" such as Heisenberg, who himself had not thought deeply enough to see that a bomb could be made, and indeed had failed to see the possibility of plutonium.

This last possibility had in fact been foreseen by one German physicist, Fritz Houtermans, who might be suspected of trying to "have it both ways"; for while he naturally wanted ultimate credit for the proposal, he told me that he tried simultaneously to establish priority and to keep it from the Nazis by putting it on record where they would never look for it—the files of the German Post Office. According to Robert Jungk in *Brighter than a Thousand Suns* (Harmondsworth, Middlesex, England: Penguin Books, 1960, pp. 93–94), Houtermans did consent to restricted publication in 1944 after he had found that Paul Harteck had independently suggested the plutonium possibility. Heisenberg himself said that the Houtermans paper was circulated (Irving, p. 278) and it seems that both Heisenberg and Carl-Friedrich von Weizsäcker each had independently foreseen the plutonium alternative (Irving, pp. 99 and 278, and Jungk, p. 91). Moreover, Irving (p. 99) quotes documentary evidence that Heisenberg, at a meeting of the Reich Research Council on 26 February 1942, conjectured that plutonium would be as explosive as Uranium-235. (Irving records that a copy of Heisenberg's talk was found to be filed under G-323 among the German papers at Oak Ridge in 1967.)

Despite this clear evidence that the plutonium possibility was known to Heisenberg, Goudsmit is hardly to be blamed for thinking that this alternative route to a bomb had escaped him. In fact, plutonium offers such obvious advantages that Charles Frank and I in British Intelligence wondered whether the apparent lack of German interest was due to someone in Germany deliberately steering his colleagues away from it. If there was any steering, though, it was more accidental than intentional, when measurements at Heidelberg, early in 1941, erroneously indicated that the diffusion length of thermal neutrons in graphite was only about thirty-five centimeters instead of the seventy centimeters that Walter Bothe had expected. This led Heisenberg to conclude that a graphite-moderated pile would not be feasible (Irving, pp. 77–78 and *Zeitschrift für Physik*, Vol. 122, 1944, pp. 749–755). It was only in 1945 that the German physicists realized the Heidelberg measurements of 1941 had been faulty; by then it was much too late to pursue the production of plutonium.

The most telling evidence which revealed what the German physicists really thought about nuclear energy in 1945 is, unfortunately, available only in part. It consists of overheard and recorded conversations between the key German physicists during their enforced stay, during the summer of 1945, in Farm Hall, a country house in England. This is how these conversations came to be recorded: My colleague Commander Eric Welsh in British Intelligence told me that the physicists had been gathered in "Dustbin," the internment center at Versailles mentioned by Goudsmit on page 123, and that an American general had said the easiest way of dealing with postwar developments in nuclear physics in Germany would be to shoot the physicists. Our resultant alarm was genuine enough for Welsh to suggest that I propose the physicists be moved to England for safe-keeping. While hostilities lasted, Farm Hall, which had been used by British Intelligence as a rendezvous for dispatching members of the various Resistance movements by air, was now free. I therefore suggested that it would be suitable. This is the answer to Goudsmit's bewilderment (p. 133) when he commented, "Just why these top German physicists were interned in England, I never understood." Almost as an afterthought, I also suggested that microphones be installed, in case the German physicists might discuss among themselves items that they might not have revealed in direct interrogation. Thus, when the news of the Hiroshima bomb came through, we were able to record their reactions. Although copies of the transcripts were sent to America and were partly revealed by General Groves, and indeed are referred to by Goudsmit, they have never been published in full, perhaps because of an official fiction that the British Government does not eavesdrop or because of a reticence to reveal, without permission, statements made by those who were unaware they were being recorded. Since I myself was responsible, on this occasion, for the recordings, and the German physicists may have considered our behavior ungentlemanly, I can at least claim to have rescued them from the risk of being shot. There is no dignity in denying our eavesdropping, especially when verbatim extracts have been published by General Groves.

Groves attached particular importance to the remarks of Heisenberg who shared the general disbelief that the Hiroshima bomb was nuclear: "I am willing to believe that it is a high pressure bomb and I don't believe that it has anything to do with uranium." And then, "There are so many possibilities but none that we know. That's certain." My own memory of the conversations is that when Heisenberg first tried to guess at the principle of the bomb he made a rather elementary mistake which led him to think that many tons of uranium would be required in the bomb itself. As I have outlined in *The Wizard War:*

> His argument, according to my memory, ran thus: it would be necessary to produce fission in an amount of uranium that would contain of the order of 10^{24} atoms (i.e. 1 followed by 24 noughts). This is a number which is about the same as 2 raised to the power of 80. Assuming that two neutrons were produced by the fission of any one nucleus, this meant that chains of 80 fissions each starting from one original nucleus would explode of the order of 2^{80} (or 10^{24}) further nuclei. Each of the last nuclei to explode would be on the average a distance away equal to a 'drunkard's walk' of 80 steps each equal to the mean distance that a neutron would travel in the uranium before striking another nucleus. Since this vital distance was thought to be a few centimetres, say 8 to 9, the final nuclei would be 8 or 9 times the square root of 80, or about 80 centimetres away. This should be the radius of the bomb, giving a mass of about 40 tons.

XV

It seemed to us as if Heisenberg was repeating a back-of-the-envelope calculation he had worked out on some previous occasion, which caused him to rule out the bomb as a practical possibility. If so, this would tend to confirm Goudsmit's contention that no attempt was made to construct a bomb in Germany because Heisenberg's junior colleagues attached too much importance to his authority; but Charles Frank also recollects that Heisenberg gave this faulty calculation as his line of thinking in 1940, and it may be worth quoting Frank's statement in full, given to me in a letter of 26 October 1967, after he had read my own account:

> As I remember it, Heisenberg gave just the calculation you quote, except that my memory says he reached the answer 5 tons—perhaps he took a shorter distance between fission events by a factor of 2. He gave this calculation at the beginning of an elegant colloquium, delivered the day after they heard about Hiroshima, in which he used a rather polished version of diffusion-and-multiplication theory (which he was no doubt familiar with from their pile work) to arrive at an answer for critical mass of the order of 1 or a few kilograms. He gave the crude and faulty calculation at the beginning, as the way they had worked it out before, and I *think* he said it was the way *he* had worked it out, and I *think* he said it was the estimate he gave at the 1940 conference about what was to be done with nuclear fission in relation to the war. I *think* he said he had done the revised calculation overnight, and I think that the whole style of the lecture implied that he was presenting a result and argument new to him and to his audience.

The revised calculation would have been consistent with the calculation he gave to the Reich Research Council on 26 February 1942 (Irving, p. 97) when he told Field Marshal Erhard Milch that the amount of U-235 required would be "about as large as a pineapple." Whether or not Heisenberg had forgotten this second calculation in his bewilderment upon learning the Hiroshima news (Goudsmit implies on page 139 that the Hiroshima bomb used plutonium, but Groves states that it used U-235—the Nagasaki bomb was the plutonium one) could only be checked by reference to the impounded recordings. Niels Bohr was positive, in conversation with me and with others, that from questions Heisenberg put to him in a special visit in autumn 1941, the Germans had been seriously considering the manufacture of a nuclear bomb; but Heisenberg said afterwards that Bohr had misunderstood him (Jungk, pp. 99–101). Similiarly, the German physicists say that their comments at Farm Hall were misunderstood or mistranslated (Irving, p. 276 and Jungk, p. 199). It is therefore all the more regrettable that the original transcripts are not available.

From the recordings, their reactions appeared to us to range over a full spectrum from Hahn's horror at the awful result of his discovery of fission to Walter Gerlach's attitude, which was that of a defeated general. Von Weizsäcker remarked, "I believe the reason we didn't do it was because all the physicists didn't want to do it, on principle." By contrast, Hahn's comment on von Weizsäcker's thesis was, "I don't believe that, but I am thankful that we did not succeed," and Erich Bagge's, "That may be so in his case, but not for all of us." (Groves, pp. 335–336).

In fact, despite von Weizsäcker's thesis, and despite Goudsmit's alternative explanation of the inhibitive influence of the Nazi regime on science, the German scientists and engineers did very creditably. On page xi of his foreword, Goudsmit suggests aerodynamics as the only branch of science in which the Germans led; but he later (p. 149) gives them credit for being "well ahead" as regards applications of science in guided missiles, the V1 engine, U-boats, and torpedoes. To this list must be added ballistic rockets, new fuels, infra-red detection, and the nerve gases.

Altogether, their achievements were remarkable, and all the more so because of the administrative difficulties that they faced. Many of these were undoubtedly due to the attitude of the Nazi regime, as Goudsmit suggests. Toward the end of the war, we in Britain were surprised that the German research and development programs had far less coherence than our own. We looked for some time for the German counterpart to Henry Tizard who was Chairman of the British Advisory Council on Scientific Policy. Werner

Osenberg, whom Goudsmit mentions on pages 187 to 201, was the nearest equivalent, but he did not come to the fore until 1943, and not to our notice until 1944, when we began to see his name on captured documents.

We had the impression that the clandestine origins of the German military research establishments before 1933 had proved a handicap. By the time they could come into the open, there were too many, each individually not strong enough to be fully effective, but too strong and independent to be amalgamated into larger and more effective establishments. This hardly seemed to matter so long as Germany was winning the war. By 1943 however, the weaknesses were apparent, and Goering—who had the overall responsibility for German research from 1940 onwards, if not before—was supplanted by Speer, who succeeded in stimulating a spate of new weapons. Our comment, though, in Air Scientific Intelligence Report No. 32, dated 29 March 1945, on *The War Organisation of German Science* (now in the Public Records Office in London) was, "We can be fairly sure that if German science could be left under the control of Speer for a generation, it would become so organised and unoriginal as to offer no threat to the outside world: but the danger that some truly devastating weapon would be discovered before then is too great for the experiment to be advisable."

Goudsmit's description of Osenberg's readiness to turn over his papers and personnel on capture and to offer his services (p. 197) rings very true and typical. I can recall a British investigator reporting to us at about this time that "no one has appreciated the meaning of the word 'collaboration' unless he has interviewed a German scientist!" But while it was easy to be critical of our German counterparts for not standing up to the Nazi regime and for their readiness to change sides on capture, we may ask ourselves what most of us would have done in their place. Churchill was generous enough to say, "Everyone is not a Pastor Niemoller or a martyr. . . . I thank God that in this island home of ours, we have never been put to the test which many of the

xviii

peoples of Europe have had to undergo." Perhaps Heisenberg's apologia to Robert Jungk in *Brighter Than a Thousand Suns* when expressing his sense of personal shame when some of his friends died in their abortive attempt on Hitler's life made on 20 July 1944 should be read in this light: "Under a dictatorship active resistance can only be practised by those who pretend to collaborate with the regime."

Goudsmit rightly draws attention to the almost unbelievably horrifying perversions practiced in the concentration camps. I can recall our feelings in 1945 upon reading the files of Dachau about which he reports on pages 207 and 208. Hopefully, his concluding chapter is headed, "It can't happen here." Nonetheless, even overlooking the McCarthy period, it is well not to be complacent: how could the concentration camps have happened in a land that produced Einstein and Beethoven?

In 1955 I was the guest, in Munich, of the German generals and radio men who had been my former principal opponents. I had delivered the "Celebratory Lecture" in the morning to an audience of some sixteen hundred, and was being fêted. One of my English friends who was present was concerned that I might be carried away by the occasion, and said, "You had better come out to Dachau with me this afternoon." We duly went; and even though it was ten years since the crematoria had last been operating, it was a very sobering experience. We returned to Munich for the evening, where I found myself at one of the tables in the Hofbräu or Loewenbräu surrounded by my German hosts, including such officers as Feldmarschall Kesselring and Generals Kammhuber (nightfighters), Galland (dayfighters) and Martini (signals). We were pledging eternal friendship when I said, "This is all very well—but I was at Dachau this afternoon. What are you going to do to stop that ever happening again?" The shocked silence was first broken by Martini, a natural gentleman, who said, "Most of us knew nothing about the concentration camps. Isn't it terrible to think how one man could have led us so far astray!" There was a chorus of support, and I believed that he, at least, did

xix

not know what had been going on in the camps. But after the chorus had died down his friend General Egon Doerstling who had introduced himself to me as a historian and a philosopher said, "It is true that most of us did not know about the concentration camps. But we cannot deny that they existed, because they are there. We do not understand how they could have happened—and because we do not understand, I cannot see how we could guarantee that they will not happen again!" I have sometimes recalled Doerstling's honest answer when contemplating trends in our own society.

In his foreword, Goudsmit writes of the Nazi prejudice against "Jewish science"and rightly says that there is no such thing. As a Jew, though, he may have been modest in not pointing out the remarkable record of the Jews in science in Germany: there could be a particular talent for science among the Jews, for although they constituted only one per cent of the total population, they gained one-third of all the Nobel Prizes awarded to German citizens (K. Mendelssohn, *The World of Walther Nernst*, New York: Macmillan, 1973, p. 28).

Two questions arise from Goudsmit's last chapter. First we may ask how can we avoid the dangers of such movements as national socialism without falling into anarchy? The second question, a related one, is what form of society is most favorable for the optimum progress of science? Democracy is generally accepted as the answer to both questions, and it is the one that Goudsmit, of course, gives. Excessive respect for authority—to which he points as one of the reasons for the German failure—undoubtedly inhibits individual enterprise and originality, particularly in the minds of younger people at their most creative period. But too little respect for authority can destroy corporate effort and lead to such destruction, both physical and moral, as did the student unrest of the 1960s. Max Born once put the point appealingly in *Physics in My-Generation* (Oxford: Pergamon, 1956, p. 108):

Complete freedom of the individual in economic behaviour is incompatible with the existence of an orderly state, and the totalitarian state is incompatible with the development of the individuum. There must exist a relation between the latitudes of freedom Δf and of regulation Δr, of the type $\Delta f \cdot \Delta r \sim p$, which allows a reasonable compromise. But what is the 'political constant' p? I must leave this to a future quantum theory of human affairs. The world which is so ready to learn the means of mass-destruction from physics, would do better to accept the message of reconciliation contained in the philosophy of complementarity.

Democracy must have enough structure and authority to accomplish its ends and due respect for that authority is essential. But this does not mean that authority should be above challenge: indeed if it cannot withstand thoughtful challenge then it ought to give way. As Churchill wrote of British errors in the U-boat war of 1917, "The firmly inculcated doctrine that an admiral's opinion was more likely to be right than a captain's, and a captain's than a commander's, did not hold good when questions entirely novel in character, requiring keen and bold minds unhampered by long routine, were under debate." If this is true of military affairs, as it certainly is, it is even more true of natural science where so many of the fundamental ideas have occurred first in the minds of young men. No leader of scientific thought, however eminent and whatever his past achievements, should therefore be above having his ideas questioned by his juniors provided that they offer him respect with the realization that a man who has been proven right in the past may have a better than average chance of being right again.

Referring to the lessons of the German example for American science, Goudsmit made the point (p. 238) that "we cannot go on forever living on borrowed scientific capital from Europe," citing such brilliant imports as Enrico Fermi, Theodor von Karman, and Stephan Timoshenko. The implied question was whether this injection of scientific capital would result in the bringing on of an American generation able to take over as the European imports (of which Goudsmit was a conspicuous member) went into retirement. But even by 1940, there were outstanding examples of American promise: I recall Niels Bohr telling me, in 1944, that Britain would have to look to its laurels in physics, mentioning particularly the brilliance of Robert Oppenheimer. And while complacency could be disastrous, and while lax trends in education may have to be resisted, the many achievements in the years since *Alsos* was first published amply confirm the plane of world leadership to which Goudsmit's generation helped to lift science in America.

R. V. JONES
Aberdeen, Scotland
September, 1982

SUPPLEMENTARY BIBLIOGRAPHY

Bagge, E., and Diebner, K. *Von der Uranspaltung bis Calder Hall.* Hamburg: Rowohlt, 1957.

Bar-Zohar, M. *The Hunt for German Scientists 1944–1960.* London: Barker, 1965.

Jones, R.V. *Most Secret War.* London: Hamish Hamilton, 1978. Also published as *The Wizard War.* New York: Coward, McCann and Geoghegan, 1978.

Table of Contents

xxiii

Foreword

THE myth of Nazi military supremacy was smashed forever by the might of the Allied armies, but the legend of German scientific superiority lingers on. More than two years have passed since the war's end, and still today not merely the man in the street, but many of our scientific and military experts as well, believe that we were engaged in a desperate race with the Germans for the secret of the atom bomb, and it was only by a miracle, by a hair's breadth, that we got there first.

To be sure, we were engaged in such a race, but that race, as we now know, was rather a one-sided affair and the situation was not nearly so desperate as we supposed. The plain fact of the matter is that the Germans were nowhere near getting the secret of the atom bomb. Indeed, at the rate they were going and the direction they were taking, it is anybody's guess if they would have arrived at it at all in any practicable period of time. Months after our scientists had established, beyond the shadow of a doubt, the feasibility of the atom bomb, the Germans were still only talking about the "uranium problem" and the possibility of constructing a "uranium machine." They did not yet know how to produce a chain reaction in a uranium pile. They did not know how to produce plutonium. They knew only, perhaps with typical German arrogance, that if they could not make an atom bomb, nobody else could either, and so they need not be unduly worried. That arrogance received a terrible jolt the day our atom bomb fell on Hiroshima.

These are some of the facts that were established beyond any possible argument by the Alsos Mission, of which the author of the present book had the privilege of being the scientific head. Appointed by the War Department at the request of Major General L. R. Groves, U. S. Army chief of the atom bomb project, it was the duty of this Mission to follow immediately in the wake of our armies in the invasion of Europe, for the purpose of determining precisely how much the Germans knew about the atom bomb and how far they had progressed in its construction. In other words, Alsos was a scientific Intelligence Mission, and since the atom bomb was TOP top secret and pretty much a mystery even to the high military echelons, our work was of necessity shrouded in secrecy. Now the story can be told.

In the following pages, I have reviewed the

circumstances leading up to the mission and have tried to tell how Alsos went about its job of scientific sleuthing and what it discovered. Quite obviously, if the only conclusion to be drawn from my narrative is that American science proved itself superior to German science, at least in the field of nuclear physics, then this book would hardly be necessary. I am interested, rather, in why German science failed where the Americans and the British succeeded, and I think the facts demonstrate pretty conclusively that science under fascism was not, and in all probability could never be, the equal of science in a democracy.

Too many of us still assume that totalitarianism gets things done where democracy only fumbles along, and that certainly in those branches of science contributing directly to the war effort the Nazis were able to cut all corners and proceed with ruthless and matchless efficiency. Nothing could be further from the truth. The only branch of science in which the Germans outdistanced us was aerodynamics and this was because Goering, who was boss of the show, cut across party lines and gave the scientists a free hand and unlimited funds. The failure of German nuclear physics can in large measure be attributed to the totalitarian climate in which it lived. There are lessons we can all learn from that failure.

Complacency, for instance, was one of the worst enemies the German scientists had. Confident of the superiority of German science, they assumed that no nation could equal them, none succeed where they failed. It would be as grave an error for us to assume that no other nation can match our country's scientific achievements simply because we were the first to have the atom bomb.

Politics, the interference of politicians in the affairs of science, and the appointment of party hacks to important administrative posts, is another grave error it would be foolish to suppose was purely a German monopoly. By putting politics first and science second, the Nazis contributed greatly to the deterioration of German scientific teaching and research. To be a good Nazi was not necessarily synonymous with being a good scientist, and we can learn from their mistakes to leave science to the scientists.

The same thing applies to dogma, whether it be political, scientific or religious. The stubborn blindness of dogma and the free inquiring spirit of science do not mix, nor is there any such thing as "Jewish science" or "Aryan science." The anti-Semitism doctrine of the Nazis was disastrous to German physics not merely, and not even principally, because it brought about the exile of certain notable scientists. These men could have been replaced in due time by younger colleagues. But by the infusion of dogma into the body of scientific thought the Nazis tended to bring the whole subject of modern physics into disrepute,

with the result that the "Jewish science" of physics became unpopular at the universities.

There was, finally, the matter of hero worship.
Werner Heisenberg, for instance, was the foremost atomic physicist in Germany, a scientist of
world repute, and no good young German scientist would think of questioning the word of the
master. But science is not authoritarian, nor can
scientific thought be dominated by a boss, however gifted he may be. Scientific research represents a collective effort. It is a matter of trial and
error, of consultation and correction on the part
of many minds. It is always possible that a
younger colleague may be on the right track and
a Heisenberg on the wrong.

To understand what the Germans did wrong
scientifically and where they failed, it is necessary for the reader to have some idea of the fun-
damentals of the uranium—i.e., atom bomb—
problem, or at least some familiarity with the
terms used in this book. Rather than interrupt
the narrative, I have placed this necessary information in an appendix. The reader may, if he
wishes, read this outline and brief glossary of essential terms at the outset, to familiarize himself
with words and phrases that must inevitably be
used again and again in a book of this sort. Or he
may wish to refer to the explanatory material at
the end of the book whenever he comes upon a
reference to the uranium problem that is not
clear to him. In any event, I have tried to make

it all as simple as possible, endeavoring not so much to explain the intricacies of the problem which would take an extended essay, as to identify its principal phases.

In his Pulitzer Prize winning book, "Scientists Against Time," James P. Baxter cites the Alsos Mission [1] as "one of the finest examples of cooperation of the scientists and the Armed Forces." It was indeed, in this respect, an ideal outfit, thanks to the superior leadership of Colonel Boris T. Pash. We had excellent backing in Washington. Colonel Charles P. Nicholas was our guiding spirit in the War Department. In the Office of Scientific Research and Development, Dr. Alan T. Waterman gave us all the help we needed. The Alsos Mission covered various fields of science but this book deals primarily with the atom bomb problem. On this subject, we had, of course, perfect cooperation from General Groves' Intelligence Department in Washington and overseas.

The enterprise was so novel, however, that it took some time before we had a clear picture of our tasks and before the mutual desire of cooperation could bear fruit. After that everything went so smoothly it would be dull to relate. I hope to be forgiven for having included a few incidents from the earlier, troublesome days of the Alsos Mission.

[1] Further information about the administrative setup of the Alsos Mission is available in "Combat Scientists," by Lincoln R. Thiesmeyer and John E. Burchard. (Boston: Little, Brown)

ALSOS

Germany's "Oak Ridge"—the Army Ordnance experimental
station for uranium bomb research near Berlin.

I

The Fear of a German Atom Bomb

LOOKING back to the end of 1942 and the early months of 1943, from the vantage point of the information we now have, one must admit that there was an element of high comedy in the state of mind that prevailed at that time among American scientists and German.

The Americans, having succeeded in producing the first chain reaction in a uranium pile, and seeing the atom bomb as a definite possibility, were certain that the Germans must know as much and more. After all, they reasoned, it was a German scientist, Otto Hahn, who had first discovered the principle of fission and another German had published the first paper on the theory of the chain reacting pile. They had begun their uranium research two years before ours. And especially, and above all, everyone knew that German science was superior to ours.

The Germans were just as positive that their science was superior to ours. To be sure, they were still groping in the dark, but, they figured, if they were in the dark where must the Americans be? Because in science, as everyone knew,

N. R. A. G. R.
Leiter d. Geschäftsl. Beirats

Geheim

9.

8. Juli 1943

Herrn
Ministerialrat Dr. G ö r n n e r t,
Stabsamt des Reichsmarschalls,

B e r l i n W. 8
Leipziger Str. 3

Rf.1481/43 g.Sr/Se.

Lieber Parteigenosse Görnnert!

In der Anlage übersende ich einen Bericht des
Bevollmächtigten für Kernphysik, Staatsrat Prof. Dr. E s a u
über den gegenwärtigen Stand der Arbeiten mit der Bitte,
dem Herrn Reichsmarschall darüber vortragen zu wollen.
Wie Sie aus dem Bericht ersehen können, sind die Arbeiten
in den wenigen Monaten doch ganz erheblich gefördert worden. Wenn auch die Arbeiten nicht in kurzer Zeit zur
Schaffung von praktisch brauchbaren Kraftmaschinen oder
Sprengstoffen führen werden, so ist auf der anderen Seite
aber auch die Sicherheit vorhanden, daß die Feindmächte
uns auf diesem Gebiet nicht mit Überraschungen aufwarten
können.

Mit besten Grüssen

Heil Hitler!
Ihr

gez. Mentzel

Letter to Goering's headquarters about the progress of the
German uranium experiments.

the Germans were way ahead of everyone else.

In July, 1943, six months after Enrico Fermi had produced the first chain reaction under the grandstand of the Chicago University stadium, the following letter was sent to the office of Hermann Goering who, in addition to his other posts, was titular head of the Reich's research council, the organization for directing and co-ordinating science in all academic institutions.

"SECRET   July 8, 1943
"Ministerial Counsel Dr. Görnnert
"Staff of the Reichsmarshal
"Berlin W 8 Leipziger Street 3
"Dear Party Brother Görnnert:
"Enclosed I send you a report by State Counsel Prof. Dr. Esau, Plenipotentiary for Nuclear Physics, on the present state of the work with the request to inform the Reichsmarshal [Goering] about this. As you can see from the report the work has progressed rather considerably in a few months. Though the work will not lead in a short time towards the production of practically useful engines or explosives, it gives on the other hand the certainty that in this field the enemy powers can not have any surprise in store for us.
"With best regards
"Heil Hitler!
"Your
"Mentzel"

5

Had this character signed his name and title in full, it would have read: SS-Brigadeführer Ministerial-Direktor Professor Doktor Rudolph Mentzel. Bearer of the "Golden Party Insignia" and a brigadier general in Himmler's notorious SS, the Elite Guard, he was administrative chief of all war research in German universities. He held a position similar to that of Harvard's President Conant, who was president of our National Defense Research Council.

But there, fortunately, the comparison ends. Mentzel was in reality only a second-rate chemist who had climbed to his high post through the devious channels of Nazi Party politics. The real German scientists referred to him contemptuously as "the culture sergeant" and joked about the time he had given his one and only lecture on chemistry. He had never realized, Mentzel remarked on that occasion, how much harder it was to deliver a scientific lecture than to make a political speech. Actually, even his Party colleagues must have looked on him with a rather fishy eye, for there is on record a secret Gestapo report about him. The report complained bitterly that his files were in such disorder it was impossible to survey and evaluate the scope of German war research and that, moreover, most of his sponsored researches seemed to be of no military value whatsoever.

But of course, back there in 1942-43, none of us knew about Mentzel or his pessimistic-optimistic report to Goering. Our scientists realized clearly the dreadful implications of the atom bomb, if it could be put together, and being men of good will many of them secretly hoped that the thing would be too difficult to achieve during the war. When they found it was not only not impossible, but highly probable, that they could make an atom bomb that would work, they became a little scared, more than a little. The thought of German superiority drove them almost to panic.

Their reasoning was logical to the point of simplicity. It must be remembered that many of them had had at least part of their scientific education at German universities, and some of the foreign-born scientists all of their education. They had a natural and, in large measure, a justified admiration for German science—before Hitler. Since the Germans had started their uranium research about two years before us, we figured they must be at least two years ahead of us. They might not have the bomb yet, but they must have had chain reacting piles going for several years. It followed that they must have fearful quantities of artificial radioactive materials available. How simple it would be for them to poison the water and food supplies of our large cities with chemically non-detectable substances and sow death wholesale among us by dreadful invisible radiations.

The fear was so real that the scientists were

even sure of the place and the date of Hitler's supposed radioactive attack. The Germans must know, they thought, that Chicago was at that time the heart of our atom bomb research. Hitler, loving dramatic action, would choose Christmas day to drop radioactive materials on that city. Some of the men on the project were so worried they sent their families to the country. The military authorities were informed and the fear spread. I heard rumors that scientific instruments were set up around Chicago to detect the radioactivity if and when the Germans attacked.

It was then and there that our atom bomb project became a "race" with Germany. Luckily, there were realistic minds at work among the scientists and the military whose actions were governed by more than fear. They decided that something would have to be done about the German "menace."

The first positive action was the destruction of the Norwegian plant for the production of heavy water, the most preferable substance to mix with uranium to obtain a chain reacting pile. This destruction, one of the minor Allied triumphs of the war, was performed by British Intelligence, the Norwegian Underground and American bombers led by Bernt Balchen.

Of course, the Germans concluded that we wanted to put a crimp in their uranium research, because we knew they were ahead of us and we were afraid of them. The blasting of the Norwegian source of their heavy water supply was a serious handicap for their research. However, they had a supply on hand sufficiently large to continue. Our raid did not stop them. On the contrary, it intensified their efforts, and at the same time it increased their self-confidence. Nor did the fact that they rebuilt the Norwegian heavy-water plant faster than was expected give us any cause for congratulations. It was a clear indication that the uranium project had a high priority in the German war effort.

That was about all we knew. We might have assumed that if the enemy's uranium development had been really on a large scale, some reliable information about its size would have leaked out via "neutral travelers" and captured soldiers. That was what happened in the case of such huge undertakings as the V-2 project. But the fact that there was no such information on their uranium project we interpreted as successful secrecy on the part of the Germans. It might just as well have been due—and in fact was due—to the insignificance of their effort. But we did not know and could not take a chance.

The same attitude on our part was revealed when we studied the German scientific publications that reached us through neutral countries. Whenever they printed something we wouldn't have published, we concluded it was done deliberately to mislead us.

French colleagues, who had escaped shortly

after the fall of France, had told us about the German interest in the famous French laboratory for nuclear physics, under the direction of the foremost French physicist, Frederic Joliot-Curie, the son-in-law of Madame Curie. We learned that a German general had come to Paris with the intention of removing all the important apparatus to Germany. Later on it was decided to leave the equipment in place and send German scientists to work at the Paris laboratory instead. This indicated a decided interest in nuclear physics on the part of the Germans. But it should also have made it clear to us that the enemy did not have adequate apparatus themselves for this highly specialized type of research.

In short, we knew very little about the German uranium project, and what little we knew we almost invariably interpreted in their favor. In the long run, this was probably all to the good, since it accelerated our own work enormously. But in those days, before the invasion of Europe, we would have given a great deal to know more.

Ordinary Intelligence information yielded nothing of value. There were always fantastic rumors floating around about terrifying secret weapons and atom bombs which were duly reported by the O.S.S. and British agents, but invariably the technical details were hopelessly nonsensical. The reason was obvious. No ordinary spy could get us the information we wanted for the simple reason that he lacked the scientific training to know what was essential. Only scientifically qualified personnel could get us that and a Mata Hari with a Ph.D. in physics is rare, even in detective fiction.

Nor did German Intelligence function any better. That they knew nothing about our work over here is clear from the letter quoted earlier in this chapter. What information they got made about as much sense as what we received about them. Reports reached them that U. S. scientists were working in large numbers on an atom bomb, but the details offered to corroborate this were scientifically so absurd that the German physicists ignored the whole thing.

If only we could get hold of a German atomic physicist, we felt, we could soon find out what the rest of them were up to. To us physicists the problem seemed very simple. Even those of us who were not working on the atom bomb project knew pretty well what was going on over here. No amount of military security could have prevented us from knowing, difficult as it was for the military to understand this. Active scientists engaged in the same general field of research inevitably form a kind of clan; they work closely together and know all about each other's specialties and whereabouts. You can't take a group of key scientists from their accustomed haunts and have them disappear in some remote place in New Mexico, together with their families, without their colleagues who are left behind wondering about it and deriving the right conclusions. The same thing, we knew, would be true of the Germans.

For this reason, some of us urged that an attempt be made to contact German scientists in Switzerland. Thus, at one time we heard that Werner Heisenberg, the foremost German atom physicist, was attending a meeting in Switzerland. Many of us had known Heisenberg for years before the war and had been very close to him. If just one of us could have a short talk with him in Zurich now we would probably learn all we wanted to know. But so would he, was the objection of the wiser military minds.

They also understood very well that as a spy a qualified scientist would be a washout. He would in all likelihood give away more secrets than he could collect. It takes more than a false beard and forged papers to keep a scientist from identifying himself to other scientists, even if he can keep his mouth shut, which is unlikely where his specialty is concerned. Not that he will give away secrets deliberately. I have been present at discussions where two men were not supposed to exchange certain pieces of information. They did their level best to keep mum and discuss only what they were allowed to. But their very silences or denials on certain topics gave away everything to the rest of us present.

These were the conditions that prevailed

among us at the time our armies were being readied for the invasion of Europe. We knew strictly nothing about the development of the German uranium project. We assumed that their progress must be parallel to ours and, in all probability, ahead of ours, and we were plenty jittery. Hitler had boasted about secret weapons. What else could he have in mind but an atom bomb? We had obtained Intelligence data on the V-1 and V-2. What final use could they be to the Germans unless they were meant to carry atomic explosives? Reconnaisance planes brought us pictures of mysterious installations along the coast of France. Might these not be bases for uranium piles to produce bombs or at least huge amounts of radioactive poisons? It was not surprising that our invasion troops were equipped with special detectors for radioactive materials. Fortunately they proved to be unnecessary.

II

We Prepare to Investigate
German Science

THE original Alsos Mission, or Urmission as the Germans might have called it, was a small unit of scientific and military personnel sent to Italy by General Groves. They had returned before the fall of Rome after collecting what material they could at the universities of Naples and southern Italy. As had been expected, their findings were not of major significance from the scientific point of view. However, the fact that such a mission was feasible, and especially their success in operating, produced results in Washington. It was decided to follow up the invasion of France with a much larger mission covering all scientific war research and sponsored by the Army, Navy, Vannevar Bush's Office of Scientific Research and Development, and General Groves' Intelligence Department.

Colonel Boris T. Pash, military chief of the first Alsos Mission, was a natural choice for the same post in the new outfit. Just why and how the author of the present book came to be chosen as the scientific chief is still somewhat of a riddle to me.

Many months later I learned from a paper which a careless secretary had filed in my dossier in Washington that the prospect under consideration—i.e., myself—had "some valuable assets, some liabilities." The liabilities, it seemed to me, were many; the assets, sufficiently strong to have me taken away from the radar work I was doing at the time, were possibly two. In the first place, although my field was atomic physics, I was not working on the atom bomb project; in other words I was expendable and if I fell into the hands of the Germans they could not hope to get any major bomb secrets out of me. In the second place, I was personally acquainted with many of the European scientists, knew their specialties, and spoke their languages.

At any rate, the assets must have outweighed the liabilities; I got the job. But it was only after I had been briefed and had passed my "screen test" in Washington that I understood fully what the job called for. I had thought it was to look into the German radar development and physics in general. I was, therefore, taken somewhat by surprise when a major on the screening committee took me aside and said, "You understand, of course, that what you are really going to do is to look into the atomic bomb development."

However, my instructions were to find out what the Germans had been doing in scientific work in general. Despite the fact that the atomic bomb people interpreted this as just so much camouflage to hide the real project of the Mission, I tried to the best of my ability, and the capacities of the Mission, to see that other fields of scientific activity were not neglected.

I was the only scientist on the Mission who was officially briefed about our own A-bomb project. The Army men were afraid I might know too much about it, and this—if I may flatter myself for the moment—may have been one of those liabilities mentioned in my dossier. I was to be accompanied by a representative of General Groves who was to keep an eye on me. This was the Major who had taken me aside to tell me what the real purpose of the Mission was. He had very powerful authorizations and dealt directly with the highest U.S. and British officials, without benefit of intermediaries.

I came to think of him as the "Mysterious Major" and at first had some difficulty making him out. I was convinced that a competent spy organization had furnished our Intelligence with detailed information on the German atom bomb and when the Major, despite my insistence, told me nothing about it I thought he was holding out on me. His silence seemed too mysterious for my liking. It was only later I discovered that most of his silence was due merely to lack of information; we simply didn't have any spy organization capable of dealing with nuclear physics. So our Major, despite his pipeline to the highest officials, was not hiding behind any cloak of mystery after all. After I realized that, we got along fine, and although he is now out of uniform and a successful architect, we are still close friends.

Shortly after we were installed in Paris, two civilians joined the Mission who had some inside knowledge of the A-bomb secrets. Fred was an engineer with a thorough knowledge of Europe, a crackajack at sleuthing and a marvellous companion. Jim, too, proved to be a tireless field operator. But the rest of the Mission was officially not in the know, although it is obvious that the physicists and chemists, in spite of their not having been briefed, nevertheless knew a lot about the uranium problem. Men like Walter Colby, physicist from Michigan, Carl Baumann, biochemist from Wisconsin, Charles P. Smyth, chemist from Princeton, Gerard Kuiper, astronomer from Yerkes Observatory, Ed Salant, physicist from New York University, Allan Bates of Westinghouse, and several of our short-term members knew more than our military security men could believe.

The original plans for Alsos seemed rather simple, viewed from behind a desk in Washington. Only a small number of scientists were to be permanently attached to the Mission; others were supposed to be sent over whenever the need arose. It never worked out that neatly. When we needed the men they were rarely there and red tape made quick trips from the U.S. to the theater of operations impossible. As it turned out, during the most important period of operations, we had only a skeleton staff.

It was the task of the scientists to obtain and analyze all pertinent information having to do with German science. From such information they had to deduce just what places, institutions, buildings, and people in enemy territory were important for giving us the information we wanted. It then became the task of Colonel Pash and his men to see that we got to these people and these places before anyone else got there. They also had to supply us with all relevant intelligence collected by other groups in the American and British armed forces.

However, the major problem that still had to be ironed out was how the co-operation between the military personnel and the civilians was going to work. This was something new in military history and there existed practically no precedent. Who was boss and when, and who decided what to do and where to go and how to do it? The Alsos Mission was later cited as "one of the finest examples of co-operation between the military and civilians." Colonel Pash never failed us. We had complete confidence in him and his men to get us anywhere we wanted, without benefit of red tape, as soon as an important objective was taken by our Armed Forces. In turn, Colonel Pash never doubted our judgment when we said that a certain "target" was of the highest importance. Alsos fully demonstrated that it was possible for civilians and military to work together on the same mission smoothly and efficiently, if their efforts were based on mutual confidence and understanding.

Colonel Pash was a master in selecting his coworkers. He had a well-rounded team of officers, who besides being splendid fellows in themselves, were all experts in administrative and operational matters. There was George Eckman, a Lieutenant Colonel, who was a wizard at cutting red tape; Dick Ham, a Major and a first-rate lawyer, whose administrative skill kept us out of trouble. Field operations were under the skillful command of Major Robert Blake or Captain Reginald Augustine. Several of the officers and enlisted men were counter-intelligence agents, and therefore must remain nameless here. They were excellently trained men, well educated and with a knowledge of one or more foreign languages. One of the Mission's officers, Major Russell A. Fisher, had been selected by me because he was a physicist, whom I had known for many years. The whole outfit—the drivers, mechanics, agents, clerks under Sergeant Lolli, our splendid photographer Micky Thurgood—exemplified the best kind of teamwork. One and all, they were devoted to Colonel Pash. It was indeed a unique organization and the personnel was well aware that they were on an important assignment and in a privileged position. We were also unusually lucky with the choice of Major Howard J. Osborn to be our man in the Pentagon, who kept our channels with Washington wide open.

The Alsos Mission was the favorite child of certain Intelligence officers in the War Department, and of General Groves' Intelligence organization in Washington and overseas. A number of scientists, mostly civilian, were attached to this heterogeneous group, a few for a long period, several for shorter assignments. It is an amazing truth that the whole outfit worked like one large family with only very minor quarrels, although naturally, there were some instructive clashes before this benign situation was realized.

We had thought at first that it would be possible to direct the operations of our teams from Paris headquarters, but the lack of adequate communications in forward areas made this impossible. The teams we sent out to different places had to act autonomously and make their own decisions. Advance planning was of little use, because there were always unexpected situations.

For example, when Allan Bates with two of our counter-intelligence agents entered the small town of Urach in Southern Germany, to investigate an evacuated metallurgical laboratory, they found the town in an uproar. Liberated drunken displaced persons had started a riot which was reaching a dangerous pitch. Bates immediately organized and armed (from a store of confiscated weapons in the town hall) a company of French prisoners of war, set up a temporary civil government, quelled the riot, acted as Mayor pro tem, established a system of food distribution—all illegally but necessary for the protection of our two agents, himself, and the German scientists to be investigated. Alsos civilians and military were fortunately the kind of men who were capable of making decisions on the spot; they required little supervising.

If it was impossible to direct the Alsos field operations from Paris, it was even more impossible to do it from Washington. The officers in General Groves' office who were responsible for this job were able and likable fellows even if they weren't scientists but just lawyers. But they were reluctant to leave us alone and so they tried to guide us by remote control, to the occasional despair of us overseas, by whom any order from Washington had to be regarded like the Ten Commandments from Moses, an absolute must, with no maybe's. Impressive-looking pink radiograms, marked TOP SECRET, and even worse, stamped ACTION COPY, could put any office in a dither. There was usually a time limit for ACTION, often a mere twenty-four hours.

Here is a typical, although admittedly extreme example of what could happen, what we never reckoned on when we were laying the groundwork for the Mission in Washington. We had collected a few bottles of water from the river Rhine as soon as that river was reached. Captain

Robert Blake had gone on the bridge under fire at that dangerous spearhead in Holland late in September, 1944, and midstream had filled the bottles with the precious river water. We figured that if the Germans had a large atom bomb plant, they might be using the Rhine or its subsidiaries for uranium pile cooling. If so, some radioactivity might be conveyed to the water and this in turn would be detected by us with super-sensitive instruments.

The bottles were wrapped carefully for shipment to Washington. At the same time, to please his colleagues back home, the "Mysterious Major" added a fine bottle of liberated French wine, and wrote on the label: "Test this for activity, too." It seemed a nice thing to do and a rather entertaining idea.

A few days later back came one of those top secret action radiograms. "Water negative. Wine shows activity. Send more. Action." Well, that was pleasant, we thought. They had appreciated our gift and were carrying our little joke further. But we were quite mistaken. Further radiograms indicated that they had actually measured the radioactivity of the wine instead of drinking it. Might it not mean that the Germans had a secret laboratory somewhere in the mountains where the wine was grown? It did not occur to them that there are many mineral waters in France which show weak radioactivities.

The wording of the radiograms indicated clearly that they had been sent by someone not acquainted with elementary physics. I could not spare anyone to look into this wine nonsense. We were very busy preparing for the fall of Strasbourg, which was expected momentarily, and our whole Mission was reduced to three physicists and a few experts in other fields. I refused to act unless I received a radiogram with intelligible information signed by a reputable scientist on the project.

I lost the battle. I was forced to send one of the physicists, Major Russel A. Fisher, on a ten-day wild goose chase for radioactive wine. Fortunately we knew where the wine came from; it was a bottle of excellent Roussillon which Major Fisher had liberated when he came to France on the heels of the southern invasion. Now I had to send him all the way back to the Marseilles region to find more of the same.

"Do a complete job," I said. "Don't be stingy with the confidential funds. And above all, be sure that for every bottle of wine you locate, you secure a file copy for our office in Paris."

Major Fisher, accompanied by Captain Wallace Ryan, had an interesting time. They had no trouble at all in getting all the wine and all the information they wanted, free of charge. In fact, the French wine dealers thought that these two officers were American business men using, or rather misusing, their Army status to reopen relations with French exporters. Everywhere they were received with open arms and urged to sample more of the delicious liquid than they were used to. But everything went without mishap. They returned to Paris with a most representative supply of Rhône wines, samples of the grapes, the soil where they grew, the water of little rivers, wholesale samples, retail samples, all absolutely complete. Except for our file copies, all of this was sent on to Washington accompanied by Major Fisher's report. Perhaps his unusual task and its bibulous hardships inspired the Major, for certainly his report was one to end all Intelligence reports.

As time went on, remote control of Alsos was reduced to a minimum and whenever necessary we used the wine operation as a strong argument for being left alone. We never did find out what happened to the wine we sent over. We hoped that the recipients had used it the way nature intended wine to be used, and had not wasted it on test tubes and retorts.

There was one other occurrence that helped clarify the problem of freedom for the Alsos Mission. It had been hard to convince our military Intelligence colleagues that searching for scientists was not the same as searching for spies or criminals. A list of "suspects" had been furnished us. In military jargon they were called "targets" and our job, in the normal police procedure, would have been to investigate each one individually.

It was hard, at first, to convince the military that not all these "targets" were worth searching for, that some of these scientists were important and others utterly unimportant. They could not understand how we could know, in advance, just which of the enemy scientists had the information we wanted, although any reputable scientist working in the same field would have known the same thing.

One day we received a dispatch from Washington about a mysterious German scientist who had visited the United States before the war. Instead of furnishing us with data about his scientific work, which would have told us everything or at least aroused our interest, the secret message told us about his beer drinking habits, his opinion of American women, how he had had German measles in 1938, had a polyp in his right nostril and an atrophied left testicle. After this memorable dossier, even the military members of our mission treated such dispatches from Washington with some skepticism.

The wine incident had a sequel, however. General Groves' scientific advisor reprimanded me later, justly and severely, because I had failed to see the danger of playing a practical joke in wartime, across three thousand miles of ocean, and on such a vital subject as the atom bomb project. Washington could take no chances.

III

The Need for Secrecy

SECRECY, it was impressed upon us at the outset, was imperative, despite the fact that our code name, ALSOS, seemed a give-away, being the Greek translation for Groves. Since General Groves was in complete charge of all Army activities relating to the atom bomb project, the inference did not take too much imagination. To make it even more obvious, the Mission's vehicles had license plates bearing the Greek letter Alpha.

That the enemy wasn't supposed to learn anything from us about our atom bomb work was obvious. They couldn't have learned the "know how" from members of the Mission anyway. But our own people were to be kept in the dark, too. The Alsos Mission would have to operate without telling anybody what we were interested in. Only a few top U.S. and British officials were informed about our task. As I have already pointed out, there were officers and civilians attached to the Mission who were not supposed to know what our highest priority interests were. Furthermore, there were already in existence Army, Navy, and Industry investigating groups trying to find out something about German scientific work. It became one of the Mission's more painful tasks to keep these unauthorized sleuths out of the running, without giving away our own identity.

Thus in November, 1944, I received a call at our headquarters in Paris from Joliot-Curie. He reported that a couple of officers direct from Washington had called on him and asked him to tell them all he knew about atom bomb research. "I have already told everything I know," he had replied. "Don't you realize you have a special team here for just that purpose, the Alsos Mission?" And he proceeded to give them our address. Fortunately, the men in question never showed up, proper measures having been taken to lead them in another direction. But after this revelation, I proposed to my colleagues that we put out a shingle—Alsos Mission, Atom Bomb Disposal Squad.

Again, in September, 1945, after Hiroshima, the Mission received a long distance call from the principal office of U.S. Military Intelligence in Germany. They had located the laboratory of Otto Hahn, discoverer of uranium fission, the basic process of the A-bomb, and should they pick him up for us. When I told them we were not in the least interested, they seemed quite amazed. They would probably have been even more amazed if I had told them we had called on Dr. Hahn some five months before and that he had been interned since then.

Of course, even though very few knew specifically what we were doing, almost everybody knew why we were on the Continent; and everybody knew it was top secret. This gave us an aura of undeserved importance which greatly facilitated our work and provided us with unheard of and sometimes unnecessary privileges. Whenever we needed some special assistance or favor, we would whisper into the ear of the proper sergeant or general, Atom Bomb! Invariably, it worked wonders and proved more efficacious than any official papers from London or Washington. The importance and secrecy of the whole thing caught the imagination of outsiders and they were eager to assist us.

This extreme and even highly imaginative impressionability sometimes had amusing consequences. For instance, we had obtained some German cubes of uranium metal. They made excellent paper weights. A high ranking officer who visited us was so impressed by this display of top secret stuff that he could not refrain from talking about it. It was not long before we received strict orders from General Groves' office to hide those uranium cubes in our desks. It was too bad. They made such nice paper weights.

It was naturally assumed that the Germans would leave spies and informers behind when they retreated; we were warned to be on the lookout for them. In vain. There weren't any, at least none who thought our outfit was worth their trouble. We were also told to watch out for the local personnel we used, drivers, cleaners, hotel employees, and the like. The nearest we got to a suspicious character was a hotel porter who thanked me for a tip with "danke sehr." But that was only a question of habit. Poor fellow! He had been saying it for the past four years and the Germans had only left a couple of days before.

As a matter of fact, the Germans never did find out about us, except for some of their top scientists, and that was only after they had been taken into custody and questioned. Even then their only assumption was that they knew more than we did, that they had in their possession atomic secrets we still were ignorant of, and that was why we had arrested them for interrogation. Hiroshima took them all by complete surprise. Only once, when we captured four nuclear scientists in Strasbourg in November, 1944, did word get back to Germany about us through a laboratory assistant. But she did not know our names, nor what we were really up to, so her information was not particularly useful to them.

Of course, we did not have a corner on the secrecy business. The Germans had their top secret, too, although they were less careful about it than we were. They could not refrain from some imposing window display. One of the first clues we found in Strasbourg was an envelope with the imprint: The Representative of the

Reichsmarshal for Nuclear Physics. This gave us quite a shock; it looked as if the Germans had a full Marshal in charge of nuclear physics, whereas we had only a two-star General. But then it dawned upon us that this was just one of those German language pitfalls; what was really meant was "The Representative for Nuclear Physics of Reichsmarshal Goering." We found out later that the physicist who held this important post was dubbed "Reichsmarshal for Nuclear Physics" by his colleagues. But an envelope like this revealed to the general public the importance which the government attached to this branch of science and even a second-rate spy could have connected this with atom bomb work. The same was true for the blanks filled out for the use of government vehicles, trip tickets we would call them, which bore the same imprint as the envelope and the further statement, "The trip is of importance for the war effort."

It is true that these envelopes, trip tickets, and further elaborate letterheads were used only in Germany proper. But our security authorities would have thrown a fit if we had done the same thing here. If we had gotten hold of one of these papers early in the war, our fear of a German atom bomb would have been intensified. But we never knew anything reliable about it until we knew practically everything at once and that was after the fall of Strasbourg.

There are still a few secrets which members of the Mission are not supposed to reveal. We are not supposed to tell just who among the Army personnel were directly connected with the A-bomb Intelligence. We cannot divulge how much uranium and heavy water was found in Germany and what was done with it. We helped find it, but never knew how much it was until later press dispatches from Germany told us about it. We don't know what became of the stuff, except for the few souvenir pieces we kept. Alsos scientists sometimes resented not being informed more fully. We could not see why Alsos officers were better qualified to use certain secret information than scientific experts. I guess now that in most cases the Alsos military did not officially know so much more either, but their grapevine worked better.

We are often asked just how we managed to obtain the German secrets. It must indeed look like a formidable achievement to the outsider. There was all of Germany to cover, and a handful of men with jeeps were supposed to find, at the earliest possible moment, the hidden laboratories, their scientific staff, equipment, material, and their secret papers. That we accomplished this sounds impressive, but in reality it was not. Our secret method of operation was like the so-called secret of the atom bomb itself. An algebra problem can be a deep mystery to the uninitiated and duck-soup to a high-school kid.

Nevertheless, we did occasionally impress the more seasoned professional Intelligence operators. How did we do it? How did we know just where to go? How did we know who had the secrets? How did we know who was important? The last question is probably the key to the whole matter. To an outsider, a professor is a professor, but we knew that no one but Professor Heisenberg could be the brains of a German uranium project and every physicist throughout the world knew that.

There are people who ask us every so often, whether we are absolutely sure we now know everything the Germans did. How can we be sure that somewhere in Germany, still hidden, there isn't a group of men, whom we have never heard of, secretly manufacturing atom bombs even now. There were even Intelligence reports referring to such a possibility. During the time the Russians occupied the Danish island of Bornholm, one heard frequent official and unofficial rumors to the effect that there was a group of German scientists on the island who had completed an atom bomb. We came across similar rumors frequently during our investigations.

I still do not know how to explain the absurdity of these rumors and how to convince non-scientists. Possibly a paper hanger can become a military expert, and a wine merchant a diplomat; but an outsider simply can't acquire the necessary scientific knowledge for making an atom bomb overnight. We are always told, with some exaggeration, that only a dozen people in the world understand Einstein. It follows that at least one of that dozen must be included in any atom bomb project since its construction is closely tied up with Einstein's theory! In other words, we knew who our chief targets were in Germany before we started. What we had to find out was how far they had advanced on their atom bomb project.

IV

Operation Cellastic

COLONEL PASH entered Paris with the very first Allied troops. We civilians came two days later. Our first contact was naturally with the foremost French nuclear physicist, Joliot-Curie. His laboratory at the Collège de France had been making "Molotov cocktails" and other home-made explosives for the French resistance and the guard we posted there was scared stiff. He, like most of our military personnel, was a counter-intelligence agent, and spending several days and nights around a dump of home-made explosives was not his idea of counter-intelligence work. He had a decidedly uncomfortable time of it until the experts arrived to take the stuff away.

Joliot told us all he knew, but after several conversations it was plain that he knew nothing of what was going on in Germany. He confirmed our previous information that two German high officials had visited him immediately after the occupation, a Professor Erich Schumann and a Dr. Diebner. They wanted to remove the cyclotron and all other scientific equipment to their home laboratories. But later there was a change of plans and instead they sent physicists to work in the Paris laboratory.

It was the fact that German scientists were known to be working in his laboratory that gave rise to the rumors, current during the German occupation, that Joliot was a collaborationist. As it turned out, Joliot had very little to do with the Germans who had taken over his laboratory. Far from being a collaborationist, he was involved in the French underground resistance movement. The fortunate and truly amazing thing was that Gentner, the chief of the German scientists in Paris and the only one with whom Joliot was on friendly terms, knew about the French scientist's political activities and protected him from the Gestapo. Gentner himself was an anti-Nazi who had worked in the United States with Ernest Lawrence, Nobel Prize winner and inventor of the cyclotron. A sound scientist and a man of good will, Gentner himself was under suspicion because he was not a Nazi and because, according to a secret Gestapo report on him we later uncovered, he was thought to be under the influence of his wife, who is Swiss.

Recalled to Germany around the end of 1943, Gentner got word to Joliot that the man who was being sent to take his place was a real Nazi and to be wary of him.

This was all very interesting, but it did not

35

advance us any in our search for the enemy's A-bomb. The few letters the Germans had left behind in Joliot's laboratory told us nothing. One of them, for instance, was from a lady friend requesting its recipient to bring back some Chanel No. 5 when he returned. The directions were very German or maybe just very feminine. The money she had sent was for cigarettes, tea, and perfume; if there wasn't enough money, then cancel the cigarettes. If there still wasn't enough, then never mind the tea. But the Chanel No. 5—under any circumstances.

This secret missive, needless to say, lay outside the scope of Alsos. Paris, for our purposes, was definitely a disappointment. There was nothing much we could do but wait out the interval until the front had moved closer to our objective.

More to keep up our morale than in the hope of finding anything concrete in the way of clues, we "raided" the now deserted Paris offices of various German industrial concerns. Siemens particularly interested us. They were the biggest German electrical concern and we knew that their research chief was an excellent physicist. We knew, too, that he had been designing cyclotrons and was probably working on the uranium project.

Siemens was located on the top floor of a walk-up apartment house. Since nobody had been there before us, we found some nice electric heaters and some good office equipment, which we promptly confiscated. But we discovered little else of value to us, despite the fact that we followed up the names and addresses we found in several stenographic notebooks. One of these names was a Mademoiselle Carola and we thought maybe—at last—we had discovered a genuine Mata Hari. But our Mademoiselle Carola turned out to be only an aged seamstress.

Then one day our high level contact with Supreme Headquarters Intelligence, Sergeant Simon of the SHAEF Document Center, delivered a stack of German papers for our perusal. They were mostly bills from French manufacturers of routine electrical equipment delivered to a German firm in Paris. But one of the bills attracted our attention, it was obviously not for common material but for a development model of some radar equipment. We soon located the French engineer who had done this work, hoping that he could tell us just what important enemy agency had sponsored his research and who the personalities were. The man was quite co-operative, especially after a couple of good U. S. dinners and lots of French wine, but he was unable to tell us what we wanted.

In such discussions and interrogations, one cannot help revealing one's interest and so the French engineer said: "I see you are a scientific Intelligence group. Do you know that during the occupation the Germans had a similar organization here in Paris! I worked for them. All I had to do was to keep them occasionally informed about new inventions and inventors. They called their organization Cellastic and used some Dutch and Swiss people who feigned an interest in buying patents as a front. But it was really a German technical spy ring."

He gave us the address, 20 rue Quentin Beauchart, just off the Champs Elysées. It was right next to the headquarters of the American cloak and dagger organization, the O.S.S. We drove down at once. It was a little palace and neighbors told us that it belonged to descendants of Napoleon's Marshal Ney. Before the war it had been leased to the Legation of Venezuela. We had no trouble entering, and as was our custom in those cold October days, we first paid a visit to the wine cellar. There was no wine at all, but hundreds of empty bottles, which were if anything scarcer than wine. These we later presented to the officer-in-charge of our billet, so that wine could be bottled in the country and distributed among the officers.

The place looked very empty, as if they had had ample time to evacuate. They must finally have hurried, for downstairs in the large recreation hall, crockery and glasses were left behind in the process of being packed. The place looked queer, some rooms had been sound-proofed and there were interphones between the offices of a special type that could not be tapped. Upstairs were the remnants of a primitive chemical laboratory.

We began to scrutinize the place more closely, for we knew that it is absolutely impossible to remove all clues. What was left of the library indicated indeed that the former occupants had a strong interest in pure and applied science. Soon we found more important clues such as a discarded floor plan with the names and technical interests of the occupant of each office. A carbon paper gave the names and addresses of all French employees; the telephone switchboard yielded a list of messages and call numbers for the last two months. The doorman's list named all visitors with dates and times over several weeks.

This find was a great shock to our "Mysterious Major," who suddenly realized how easy it would be for a spy in Washington to obtain such information. "My God!" he exclaimed, as one who suddenly sees the light, "Washington is full of such lists. Anybody can see who visits Vannevar Bush, or Geegee." (General Groves)

Finally, we found desk calendars with entries in Dutch indicating visits to the laboratories of the Sorbonne, and contacts with some of the foremost French scientists. These calendars also revealed that two prominent Dutch physicists, with whom I was acquainted, were connected with this outfit. In addition, two other young

Dutch physicists were members of the regular office staff.

Searching further, we found more clues and leads, all this in a place which at first seemed totally empty. There was unopened mail in the mailbox, a list of payments to roving employees, including our French engineer who evidently had spoken the truth in every detail.

From that time on I would always leaf through desk calendars in any office where they kept me waiting. This was especially interesting in French places, where they used the German desk pads, beginning where the Germans had left off, about September 1, 1944.

The next step seemed clear, and that was to call on those French employees whose names and addresses we had found. Judging by their pay, however, only two seemed of any possible importance, a radio engineer, and a chief secretary. We didn't have much luck. The radio engineer was in a special jail for collaborators and could be visited only by very special permission, which arrived too late. We examined his dossier in the famous Deuxième Bureau and found that he had belonged to the French Nazi party, but he had told nothing about that mysterious German office where he was employed. And the Mademoiselle had moved from room to room. Everywhere we were told how beautiful and elegant she was, which, of course, made us even more anxious to find her. Would we really come across a Mata Hari this time? Perhaps a minor one but maybe just as beautiful. The last address we tried, we were told that she had been called for frequently by a gentleman in a jeep américaine, and was not expected back. We appeared to be out of luck.

The thing to do now was to look for the scientists whose names we had found. With this in mind, Fred and I visited a well-known book dealer and publisher of scientific books, near the Sorbonne. His store was dark and cold and made a fine setting for a detective thriller. Mr. Frey, the dealer, was a Mexican citizen of European and Mexican Indian descent, who had lived in Paris for many years. He told us about his interesting experiences during the occupation and, of course, he remembered the foreign scientists who never failed to visit his store.

"Did any Hollanders ever come to Paris during the occupation?" I asked. "Very few," Mr. Frey said. "But there was one young physicist who came here frequently. I sent him some books recently but they came back as undelivered." He called his clerk, who in turn produced a package, mailed to Mr. Zwart at our place in the rue Quentin Beauchart. We were on the right trail. Mr. Zwart was one of the young Dutch employees.

"Oh, I almost forgot," our host suddenly exclaimed. "I got a letter from him the other day. He's a very nice fellow. Let me see if I can find it. I did not read it carefully and forgot what it was about."

The letter was found. It revealed that our man Zwart was now living in a small village, Pont Saint Pierre, near Rouen, with distant relatives. That was exactly what we wanted to know, but we could not leave without showing that we were motivated by more than mere curiosity and so we stayed to listen to several more tales of the occupation, while the shop seemed to get colder and colder. At last we got away, had a few drinks at the Ritz bar to get warm and made plans to visit the village near Rouen the next day.

With the assistance of the local head of the F.F.I., the French underground army, we readily located the place. We rang the bell and the door was opened by a young man I recognized at once as a Dutch student. It is minor details such as clothing, haircut, and facial expressions which reveal much to an observer and one can very often determine the nationality of strangers that way. Germans have a preference for green and brown apparel and hats of rough felt. Hollanders can often be recognized by their careless posture. One might, of course, be mistaken, but this young man who opened the door did not fit into the picture of a small French village.

"You are Mr. Zwart," I said with my foot in the door. "May we come in? We want to talk to you." When we told him what we were after, he refused to talk. Even after I let slip that we knew a lot already about that mysterious firm and mentioned some names, all he was willing to say was "I felt all along that there was something wrong with this firm 'Cellastic,' but I had nothing to do with it. I do not want others who are equally innocent to be involved, that is why I do not wish to tell you any more." I finally lost my patience. A Hollander by birth, I could no longer be polite to a Dutch collaborator, a quisling, a man from my own university, the University of Leyden.

I dismissed my companions and when I was alone with him I suddenly burst out in Dutch and told him just what I thought of his actions, his treason, his cowardice, his selfishness. He did not know who I was, but my unexpected attack in Dutch hit home. His whole being reflected fear and guilt.

I was perhaps as broken up by this experience as he was. Meeting a traitor, even a mild one, is so much worse and so much more incomprehensible than meeting an enemy. I was depressed all the way back to Paris.

The next morning we informed the Netherlands Military Attaché in Paris about our findings and turned all data on the Dutch citizens involved over to him. He summoned Mr. Zwart who by then had become quite docile and cooperative and gave us all the information he had about the outfit. Later, after the breakthrough, we rounded up some of the Germans involved in this scientific spy ring.

An unscrupulous Dutchman named Kleiter, who dealt in patents, had offered the Germans his firm Cellastic as a front for technical spying. At the head of the Paris office was a man named Ruschevej, now hiding in the little principality of Liechtenstein in the Alps. Among the Germans was a well-known chemist Professor Criegee of the University of Karlsruhe, who later told us that he was assigned to the Paris Cellastic office as an active member of the German Army Intelligence (Abwehr). Their task was first of all by means of personal contact to keep an eye upon French scientists to see that they did not work against Germany, and to make sure that any valuable discovery they made would at once be available to the German war effort. A young Roumanian woman physicist, Mlle. Tenescu, at the Sorbonne, was paid quite well for doing nothing else but phone the Cellastic crew about distinguished visitors at the University and important lectures delivered there. It was her task, too, to introduce Cellastic associates to French scientists. It was her calls which were most prominent on the phone operator's call list we found.

We interrogated her at length, but she claimed to have done nothing wrong. She was stranded without income when the war broke out and it was a very easy way to earn good money just to phone in some information, most of which could be read on the bulletin boards anyway. She claimed to have known nothing about the firm itself, hardly remembered having visited their office once. I don't believe that a smart girl like Mlle. Tenescu, never got suspicious, but then, she had to make a living some way.

The two famous Dutch scientists Professor X of Leyden and Professor Y of Amsterdam, whose names we had found in Paris, knew all the time the true character of the Cellastic firm. In fact, the Director, Kleiter, was generally suspected of being a German spy even before the war. But the underpaid professors needed money and thought they could easily outsmart the Cellastic operators. Professor X and his wife made use of the extraordinary travel privileges the Cellastic connection furnished them at a time when only high officials were allowed to cross frontiers. They eventually escaped to London via Switzerland but failed to inform the Dutch authorities of the Cellastic organization until finally confronted with Alsos evidence. There is little doubt that stupidity rather than disloyalty was the principal factor in the behavior of the Professors. "But," said a seasoned Alsos Intelligence officer and lawyer, "most crimes are committed out of stupidity rather than evil intent."

What I cannot forgive Professors X and Y, however, is that they were directly responsible for giving Cellastic two of their young and green assistants as employees, our Mr. Zwart and a Dr. K.

Zwart admitted that he had had the time of his life in Paris, the dream city of most Dutch boys as it is of young and old of other nations. He had excellent pay, a car in wartime, little work and much freedom. But that he was well on the downgrade could be seen from the lies on his French business card, where he called himself illegally "Dr." Zwart and pretended an official connection with the University of Leyden's "Chemistry Laboratory of Professor Van Arkel," one of the loyal men of the Dutch underground resistance.

A year later I paid a short visit to Holland, mainly in connection with this same Cellastic affair. I found that in Dutch university circles the case was taken more seriously than I had expected. Dutch military intelligence had brought more evidence to light after they raided the offices of the company in Holland.

My trip gave me the chance to visit the house of my parents in the Hague, where I had been brought up and where I lived all during my high school and college days. Driving my jeep through the maze of familiar streets that seemed somehow to have become smaller and narrower than I remembered, I dreamed that I would find my aged parents at home waiting for me just as I had last seen them. Only I knew it was a dream. In March, 1943, I had received a farewell letter from my mother and father bearing the address of a Nazi concentration camp. It had reached me through Portugal. It was the last letter I had ever received from them or ever would.

The house was still standing but as I drew near to it I noticed that all the windows were gone. Parking my jeep around the corner so as to avoid attention, I climbed through one of the empty windows. The place was a shambles. Everything that could possibly be burned had been taken away by the Hollanders themselves to use as fuel that last cold winter of the occupation. The stairways had been torn down, doors ripped out, parts of the ceiling, the walls—anything and everything that was combustible. But the framework still stood.

Climbing into the little room where I had spent so many hours of my life, I found a few scattered papers, among them my high-school report cards that my parents had saved so carefully through all these years. If I closed my eyes I could see the house as it used to look thirty years ago. Here was the glassed-in porch which was my mother's favorite breakfast nook. There was the corner where the piano always stood. Over there had been my book case. What had happened to the many books I had left behind?

47

The little garden in back of the house looked sadly neglected. Only the lilac tree was still standing.

As I stood there in that wreck that had once been my home I was gripped by that shattering emotion all of us have felt who have lost family and relatives and friends at the hands of the murderous Nazis—a terrible feeling of guilt. Maybe I could have saved them. After all, my parents already had their American visas. Everything had been prepared; all was in readiness. It was just four days before the invasion of the Netherlands that they had received their final papers to come to the United States.

It was too late. If I had hurried a little more, if I had not put off one visit to the Immigration Office for one week, if I had written those necessary letters a little faster, surely I could have rescued them from the Nazis in time. Now I wept for the heavy feeling of guilt in me. I have learned since that mine was an emotion shared by many who lost their nearest and dearest to the Nazis. Alas! My parents were only two among the four million victims taken in filthy, jampacked cattle trains to the concentration camps from which it was never intended they were to return.

The world has always admired the Germans so much for their orderliness. They are so systematic; they have such a sense of correctness. That is why they kept such precise records of their evil deeds, which we later found in their proper files in Germany. And that is why I know the precise date my father and my blind mother were put to death in the gas chamber. It was my father's seventieth birthday.

V

Operation Toothpaste

In Paris the O.S.S. kept us informed whenever
they dug up something they thought might be
of scientific interest. It was through their co-
operation that we got a chance to interrogate a
German airplane expert. As it turned out, he
was not really an expert; he had been a test
pilot and during the war was the representative
in France of a well-known German plane manufacturer.

The O.S.S. people had interned him in a beautiful apartment. It had belonged to a French
collaborator, but he didn't need it just now, since
he was in jail. This airplane fellow was quite
friendly and co-operative, but he did not have
any worthwhile information for us. He must
have noticed the disappointment on the faces of
the Mysterious Major and myself, for he said:
"You seem to be interested in scientific data. I
am sorry, I cannot help you; I know very little
about it myself. But in the French internment
camp I met a man named Rosenstein who has
just the information you want. He is a real

scientist and he told me wonderful things about chemistry and atom bombs."

Our O.S.S. friends promised to produce this Rosenstein as soon as possible. We could hardly wait, but our patience was well rewarded.

Rosenstein was a tallish, thin-looking man of about thirty-five, rather shy at first and over-polite—in a typical German way. He spoke French without a trace of an accent and also spoke some English. During the months that we saw him his English improved at a terrific rate. He was indeed a typical German applied scientist, with an encyclopedic amount of factual knowledge of chemistry and physics, and especially chemical technology. He carried notebooks with him which were perfectly organized, incredibly clearly written and full of information on certain fields of applied chemistry. They contained some useful data on chemical processes which were not known in such precise detail to the Allies.

Rosenstein was of Jewish descent. He had at first fled to Switzerland, where he had had a good position in industry but complained about lack of recognition for his scientific contributions and resulting patent quarrels. His family and his mind were still in Germany. His chemical knowledge was useful to the Germans and during the war they induced him to return. They promised that he would be treated as a "W.W.J."—an "economically valuable Jew"—

or even as an honorary Aryan. He fell for this bait and for a short while the Germans kept their word. He worked on poison gas protectives and held an important supervisory position at the principal plant for the manufacture of synthetic gasoline. He had the advantage of good friends among his former classmates and faculty colleagues at some of the universities, a few of whom had risen to important positions in German chemical organizations.

Nevertheless he ran into difficulties. Fellow workers in the factory, those same Germans who now claim that they never were Nazis, objected to a Jew having such an important position. A compromise failed; Rosenstein was squeezed out. The company sent him to Berlin to work in solitude on a research problem at the Engineering School. But his promised privileges never were realized; he had to wear the "yellow star" and keep out of most places. He finally made up his mind to disappear.

Rosenstein got himself some false passports and identity papers and one day he skipped out of Berlin. In Alsace he was hidden by an employee of the Swiss company for which he had worked. Then he was aided by a German officer, an Alsatian with strong French sympathies, who brought him to Paris. There he lived as a French chemist under an assumed French name and with false papers until the liberation. He tried, so he said, to give himself up to the American

authorities but was ill-advised as just how to do
it and had put it off for a few days. While he was
asking advice from a friend in a French café, he
was overheard by an agent of the French Sûreté
and promptly interned. And that is how the
Alsos Mission finally got hold of him.

Rosenstein was most valuable to us, although
only a few of the things he knew had any real
Intelligence value. He told about a gas the Germans planned to use which would cause knock
in airplane engines and subsequent breakdown,
but that the work was abandoned because the
gas might also affect the crews and that would
mean "poison gas warfare." He revealed several
more items of a similar nature. His great value
to us was his thorough knowledge of anything
in connection with German technology, his
knowledge of the country, his insight into Nazi
mentality, his acquaintance with the Nazi dis-
tortions of the German language, the numerous
wartime notations, abbreviations, names of organizations and slang expressions. He also could
read German shorthand and taught some Alsos
members French and German.

Rosenstein thus became one of the most valuable assets of our Mission. Every morning he was
brought to the office in a jeep by an armed counter-intelligence agent and at night we took him
back to his place of internment. There, too, he
was of great use to the officers as an interpreter,
translator, and handy man, and had the run of

the establishment. No wonder we had some difficulty tying him down to one room in the Alsos Mission headquarters.

Though we tried to keep him in the dark, he was far too intelligent a man not to guess what we were after. Often he was able to supplement our information with his first-hand knowledge of the subject or the location. When we found documents showing that a Strasbourg laboratory had been evacuated to a schoolhouse in a little town, it was just the school Rosenstein had attended as a boy. When a couple of Navy captains had to investigate some German chemical projects, Rosenstein briefed them for their mission. It was indeed a strange sight to see a German prisoner lecture to Allied officers. He did it very well and even wrote some lucid Intelligence reports for the Alsos Mission on subjects with which he was thoroughly familiar, such as synthetic gasoline.

The man was very happy, he was well treated, felt that he did useful work and had no further worries or responsibilities. But with the progress of our information I finally had to get rid of him, otherwise he might have learned more about our work than any one member of the Mission and soon we expected to move on. It was sad to lose such a useful man, but necessary. The only respectable way we could deliver ourselves from his presence was to set him free. Though I thoroughly resented his return to Germany from Switzerland and his short period helping the German war effort, and though at first I felt some doubt about his loyalty or reliability, I believed that his behavior during the latter part of the war and his co-operation after the liberation of Paris had earned him his freedom. Since he was officially a prisoner of the French, we informed them of our recommendation. They were very much surprised, not that we wanted to free him, but that we were willing to dispense with the services of this useful prisoner. They could not believe that we were willing to let them use him from now on. It was indeed generosity inspired by necessity.

And so it was that a short while later Rosenstein became a free man. But now the poor fellow's difficulties really began; no money, no food, no place to sleep, no work, no more access to the Alsos Mission offices. Fortunately his plight did not last long. The French authorities found a fitting position for him in their research organization. Alsos located his fiancée in Germany and his relatives, too. He got married and will probably live in Paris happily ever after. But we had a difficult time getting used to doing without this hard-working, high-precision, living chemical encyclopedia and dictionary.

Meanwhile Brussels had been liberated by the Allies—in September 1944—and since much of the world's uranium supply comes from the

Belgian Congo, we visited that city immediately. There we learned that an important chemist of the Auer-Gessellschaft, a well-known German chemical concern, had been interested chiefly in locating stocks of uranium and that he had his office in Paris.

So back to Paris we went, to the office of a French company dealing in rare chemicals and which had a monopoly on thorium. This firm— Terres-Rares—had been Jewish owned and had been taken over by Auer. Shortly before the liberation they had removed the entire supply of thorium to Germany. The same chemist who had been interested in the uranium supplies had ordered the thorium removed.

This really scared us. Thorium, we knew, could be used in an A-bomb project, when that project was in a well-advanced stage. Did this interest in thorium mean that the Germans were really ahead of us? We checked carefully what industrial uses there might be for thorium, and had to conclude that even a small fraction of the stolen amount would last ordinary industry another twenty years.

The thorium mystery became an obsession; we had to find out more about it. But the French firm could not help us. They had no idea what the Germans wanted it for. We inquired about the German personnel and were told that the chemist, though assigned to Paris, travelled a lot and was seldom there. Instead the place had been managed by a stooge named Petersen and a very capable secretary, Fräulein Wessel. This Petersen fellow, we were informed, was not too bright. On the other hand, Fräulein Wessel was bright enough for both. She was probably a Belgian, and lived somewhere in the north.

"Did they leave anything behind them when they cleared out?" I asked.

"Oh no, nothing. They cleaned out the office. Absolutely nothing."

I knew this was impossible; there are always clues. "Are you positive they didn't leave any papers behind?" I asked again.

"Really nothing. Just a couple of catalogues. But we will look around to make sure."

They brought us the German firm's catalogues they had mentioned and a few papers they had found, among which were a couple of letters of interest. These letters bore the names of the Auer chemist, Petersen, and Fräulein Wessel. They referred to long distance calls made by the Fräulein and since such calls in wartime need authorization this proved that Wessel was more than an ordinary secretary.

In addition we found a list of registered mail that had come and gone. With few exceptions the registered letters had all been addressed to the company's offices in Germany and France. But one of the last of the entries showed that Petersen had addressed a letter to Fräulein Wessel in the little town of Eupen on the German-

Belgian border. Presumably she was still there.

Meanwhile dispatches were coming in from Washington clamoring for more information about the German company and how it was involved in the thorium business. Petersen had gone back to Germany. The only thing to do was *cherchez la femme*. Eupen had just fallen into Allied hands. We would have to look for Fräulein Wessel there.

This was not a job for counter-intelligence liaison. The red tape was too long and too involved. As usual, Colonel Pash decided to do it himself. "I'll bring her back," he said, and he got into his jeep with a couple of his men and off he went.

A few days later I received a long distance call from the Colonel.

"We got him," he said. But the connection was bad, it was always bad, and I wasn't sure I had heard him right.

"Got who?" I said ungrammatically.

"Him," said the Colonel. "You know who."

"I don't know what you're talking about," I said. And then suddenly it dawned on me what the Colonel was saying. He had not only found Fräulein Wessel; he had found Petersen, too.

Petersen and his secretary were brought down to Paris; the Herr Doktor first, the Fräulein a little later. Even more remarkable, Petersen, when captured, had with him a suitcase full of documents. It was quite an event for us. Here was our own first real Alsos Mission prisoner.

58

We had great hopes that we finally had found somebody who had inside information about the German uranium project. Thus far our knowledge had been so utterly fragmentary. All we knew was that in Norway they had been working hard on the production of heavy water. In addition we had heard through neutral sources that some of the leading physicists were working in a little village in southern Germany near the family castle of the former Kaiser. Some reports had stated that there might be a secret laboratory in the Hohenzollern castle, but that idea must have been inspired by a movie thriller. I knew the castle and it was certainly unfit for scientific experiments. It seemed more likely that the German scientists were located in the nearby village of Hechingen.

We transformed our hotel suite into a tribunal for the interrogation of our prize quarry. We put on our Sunday uniforms and Colonel Pash put on as many ribbons and medals as he could find. Then Petersen was brought in. We placed him facing the window so that we could observe all his reactions, proceeded to shout dozens of questions at him. But his answers were all disappointing. Either he was hiding something, or he really didn't know what it was all about.

Greatly disappointed, I began to study the numerous papers he had in his suitcase. It was rather cold so I crawled into bed right after dinner in order to be able to study the documents comfortably and without interruption. I had not read very far before I almost fell out of bed from excitement.

This man Petersen, like all Germans, was so systematic that his papers revealed exactly where he had been and what he had done during the previous weeks. A streetcar ticket showed that he had been in Berlin only two weeks before. A hotel bill revealed that his secretary, Fräulein Wessel, had been there, too. But what excited me most was another hotel bill which showed that just before going to Berlin he had visited that little village of Hechingen near the Kaiser's castle. There was no doubt in my mind but that Petersen, on his way from Paris to the factory near Berlin, had stopped off at the German uranium research project. It seemed a tight case. It connected the theft of thorium from France with the secret project in Hechingen and with the foremost chemical firm, the Auer Company.

The next morning I could hardly wait to get hold of the fellow again. This time we pestered him with questions concerning the village of Hechingen. He claimed he had gone there merely to visit his mother. A further search of the documents showed that indeed his mother was living there, but she referred to the village in her letter as a "restricted area." So we asked him again—why was it a restricted area? Were there many new buildings there? And why so many soldiers, scientists, laboratories, so many automobiles? Our questions did not register with the man. He was at a loss to understand our interest. "Restricted area" merely meant that no further refugees from bombed cities were accepted there. In fact, his mother was ordered to move to another town. He had been there for five days and as shown by a railroad ticket, he had also visited there a year before, but he had noticed no changes except more refugees.

Once more our hopes were shattered. We weren't any further than we had been before. It seemed probable that nothing compared to our Oak Ridge existed in Hechingen, but this was not conclusive. It might have been in a neighboring town. Still it was strange that this man of the Auer Company seemed not to be aware of the uranium project at all. We studied the papers more carefully, and questioned his secretary, but all we got was merely a confirmation of what he had told us so far. He had left Paris just before the fall of the city and had been ordered back to his office in Berlin, but he had been able to boon-doggle a special trip for himself in order to visit his mother on the way.

We were not quite satisfied. Scrutinizing the papers again we found a few suspicious-looking letters which we followed up a little further, but they all turned out to be unimportant. There were documents in relation to black market dealings in radium with French doctors which we turned over to the French authorities. There was

61

a remarkable pass which he used when he transported at one time a million marks' worth of radium and platinum to Berlin. He was, according to this paper, accompanied by a German secret service official by the name of Harun Al Rashid Bey, born in Senftenberg, Lausitz province. I was convinced that Harun Al Rashid was a code name, but my more experienced intelligence officers told me that it was probably the man's real name. If they had used a code name, it would have been Schmidt. Interrogation of our prisoner confirmed this, too.

Then there was a paper with the name and particulars of a girl named Carmen von Renser. It gave her age, her address, and lots of details. It looked like a copy of a passport or an identity card. The half-Spanish, half-German name sounded as if this time we were really on the trail of a Mata Hari. When we asked our prisoner about her, he at first didn't know what we were referring to. He suddenly remembered, however, when we showed him the piece of paper. He said it was a girl with whom he had had an affair at one time in Paris and that she had charged him three thousand francs. He thought it was outrageous. "In Berlin on the Kurfürstendamm," he said, "it is only seven marks and a half per fling." So he had taken down her name and particulars and hoped eventually to inform the officials and get at least part of his money back. We didn't believe his story, and so two of our counter-intelligence agents were dispatched to find out whether they could find this girl and see if the story was true.

It took them about two days before they finally traced her. One afternoon, when I was having a conference with some outside officers about matters of policy, one of our agents—not knowing that there were visitors—suddenly stuck his head in the doorway, "Doctor, we found your prostitute," he shouted.

But two questions still remained to be answered—two questions of great importance. First of all, we wanted to know why the fellow hadn't stayed in Berlin and why we had found him so near the Belgian border. And secondly, of course, the problem we had started out with originally. What had happened to all the thorium they had taken from France?

We did some more interrogating and studied the papers more and more carefully and finally were able to put the story together, thanks to the fact that among Petersen's papers was a complete report he had submitted to his firm about the happenings in Paris during the last week before the liberation. Fräulein Wessel, with whom he was on very friendly terms, had been able to get an automobile ride to her home town. He had given her some of his belongings. He himself had been selected to drive an army car into Germany, but—as was mentioned before—he was able to travel via the little town of Hechingen to visit his mother. From there he went to Berlin where his secretary had reported, too. Then she had returned to her home town and he was sent back to the region of the French-German border to trace some railroad cars which had been lost in transit from Paris—railroad cars which were supposed to contain some of the thorium. On this trip he found himself near the little town where his secretary was living. So he thought he would spend the day with her and at the same time pick up some of his belongings. Unfortunately for him, our troops took the town before he could make his getaway. There he was duly captured by Colonel Pash.

But what of the mystery of the thorium? The documents as well as Petersen's story were amazing. The Auer Company officials, it seems, had been thinking about a long-range future, about peace time, about the time when they could no longer make money by manufacturing gas masks, carbons for searchlights, or other war materials. They didn't want the end of the war to find their business shrinking away. They realized that for a chemical company in time of peace the big money would be in cosmetics, and they had been impressed by American methods of advertising.

Now, as it happened, one of the officials had a certain patent on toothpaste—thoriated toothpaste, a toothpaste mixed with thorium oxide which was supposed to have the same effect, probably, as peroxide, and they were already dreaming of their advertising for the future. "Use toothpaste with thorium! Have sparkling, brilliant teeth—radioactive brilliance!" After all, America had its Bob Hope and Irium toothpaste. In Germany, it would be thorium, the scientific toothpaste! And to be sure that they'd be successful, they had planned to monopolize all the thorium supply they could lay their hands on. That was the real reason that the firm had stolen all the thorium from the French. It sounded incredible, but later information confirmed the story completely. And all the time we had been afraid that it was going to be used for an atomic bomb!

Though the evidence was quite conclusive, we could not accept it as significant. The fear of a German atom bomb development superior to ours still dominated our thinking, and as we had obtained no real information of their uranium project in all our investigations so far, we were still mighty uneasy. It was not until we finally came to the city of Strasbourg that our fears were dissipated.

65

VI

Operation Jackpot

AN EARLY routine visit to a French scientific optical concern had revealed that a well-known German physicist, named Fleischmann, was now a professor at the University of Strasbourg. Later, at the Philips Lamp Company in Holland, we learned that the University had ordered new equipment for nuclear research and had enlarged its staff. The first-rate theorist, Von Weizsäcker, was supposed to be there, too. At the O.S.S. office in Paris we found a recent catalogue of the University which fully confirmed our previous information. The catalogue also showed that a Professor Haagen, a leading expert on virus research, was at the University and not in Berlin, as we had supposed. If there was any important preparation for bacteriological warfare going on, Haagen would most likely know about it. Everything seemed to point to the fact that the Germans were trying to transform the French institution into a model German university and that Strasbourg was a key target for us.

The occupation of Strasbourg was postponed again and again, while impatiently we bided our time. Finally, about November 15, the occupation started. Colonel Pash went forward immediately to round up the scientists we wanted, while I remained behind in Paris to see Dr. Vannevar Bush, who had come to Europe for a short visit.

The Colonel's first telegram was disappointing. None of the men we wanted could be found. Reluctantly I had to report to Dr. Bush, just before his departure, that our Strasbourg operation, to which we had looked forward with such anticipation, was likely to be a failure. Then came a second telegram. Some of the men had been rounded up. Fred and I rushed to Strasbourg.

The reason Colonel Pash had at first failed to find our "targets" was that the nuclear laboratory was occupying a wing of the famous Strasbourg Hospital and the German physicists were trying to pass as medics. But the Colonel had soon caught up with them and when we arrived he had taken Fleischmann and his laboratory crew into custody. The latter he had interned in the Institute, but Fleischmann he had put in the local jail to prevent them from agreeing among themselves on any cooked-up story. The theoretical physicist, Von Weizsäcker, and Haagen, the virus expert, had left town some time before, and we took over Haagen's comfortable apartment for our billet and headquarters.

It was an odd sensation, visiting the German scientists we now held under arrest. These were some of the men I had been looking for, German nuclear physicists. Yet I felt uncertain of myself, and even somewhat embarrassed in their presence, especially at the prospect of calling on a colleague in jail. How was I certain he deserved to be in jail? Or was this just normal in war? The physicists in Strasbourg did not know who we were or what we were after and I did not tell them until several days later.

Our interrogation of the men yielded practically nothing, except to point out to us who were the more important among the personnel. But this we knew anyway as soon as we learned their names. We could only hope the papers we had confiscated in the laboratory and in Von Weizsäcker's office at the University would tell us more.

Tired as Fred and I were, there was nothing for it but to get at these papers at once. This we did directly after we had finished our improvised dinner of K-rations. It was a rough evening. The Germans were shelling the city from across the river; our guns were answering. Air raids and air battles raged overhead. We had no light but a few candles and a compressed gas lamp. In the center of the room, our soldiers were playing cards. Fred and I sat in a corner on easy chairs and began to scan the German files.

We both let out a yell at the same moment, for we had both found papers that suddenly raised the curtain of secrecy for us. Here, in apparently harmless communications, was hidden a wealth of secret information available to anyone who understood it. No, it was not in code. The papers were not even secret. They were just the usual gossip between colleagues, here and there a minor indiscretion, a hint, nothing really objectionable from a secrecy point of view—just ordinary memos such as we had all sent to our own friends and colleagues when we were in the U.S. Names were not even always spelled out, or only the first names were given. Obviously such notes would have told nothing to the uninitiated; on the other hand, they were not meant to be read by Allied scientists.

We found the rather indiscreet letterheads of some of the German secret atomic laboratories. The most important of these, the Kaiser Wilhelm Institute for Physics, headed by Werner Heisenberg, had been evacuated from Berlin to the little village of Hechingen. Even the precise address and telephone number were given. We wished we could fly to Switzerland and call them up from there!

We found a copy of a letter to "Lieber Werner," who was obviously Heisenberg, telling just what problems they were working on at Strasbourg. We found references to "special metal," which was obviously uranium, memos about the difficulty of obtaining the "special metal" in slabs instead of in powdered form, letters confirming our information that it was the Auer Company that produced the metal for the German experiments. We learned that "large scale" experiments were being performed at an Army proving ground near Berlin. We found parts of computations which clearly applied to the theory of a uranium pile.

Fleischmann had attended many secret meetings and taken notes in shorthand, which we soon learned to decipher. In fact, he had collected notes on many other things, such as conversations with officials on organization problems or quarrels with and about colleagues. He was a great collector of gossip and wrote it all down. Even after he was interned by Colonel Pash, he did not fail to enter every detailed happening in his pocket notebook. And all these memos, notes, computations were dated. They mentioned the whereabouts of the most prominent German nuclear physicists and the problems they were working on.

It is true that no precise information was given in these documents, but there was far more than enough to get a view of the whole German uranium project. We studied the papers by candlelight for two days and nights until our eyes began to hurt. Later, we took them back to Paris and analyzed them again in detail. Finally they were sent to Washington where they were once more analyzed, indexed, translated and interpreted. The conclusions were unmistakable. The evidence at hand proved definitely that Germany had no atom bomb and was not likely to have one in any reasonable time. There was no need to fear any attack either from an atomic explosive or from radioactive poisons.

The seemingly innocent papers we had found at Strasbourg showed that the Germans had been unsuccessful in their attempts to separate U-235. They had probably started separation on a very small scale by means of a centrifuge and they were working hard to construct a uranium pile. But apparently they had only recently succeeded in manufacturing uranium metal.

Their computation showed that as of August, 1944, their pile work was still in a very early state. They had not yet succeeded in producing a chain reaction. Their preliminary experiments had not even given them hints about certain major difficulties they would have to overcome before the pile would function. In short, they were about as far as we were in 1940, before we had begun any large-scale efforts on the atom bomb at all. Although it was evident from the papers that the work had a high priority and that the Army was taking part in it, nowhere did we find evidence of a large effort. As far as the German scientists were concerned, the whole thing was still on an academic scale.

It was in a jubilant mood that, after four days in Strasbourg, we returned to Paris, taking Fleischmann and three of his colleagues with us to turn them over to the proper authorities for internment. Fred and I kept Fleischmann in our own car, still·holding him apart from the others in the hope that we might get some more information out of him. We were unsuccessful. Our German prisoners continued to be rather defiant. They had no realization that the war was already lost, and they ascribed our special interest in them to their supposed vastly important work in nuclear physics. The situation was indeed ironic, but we were to meet with it again and again in our encounters with German scientists. To the very end they remained serenely confident of the superiority of German science. But we of the Alsos Mission were now considerably less confident of this and there were moments when it was hard to keep the joke to ourselves.

But right now the men from Strasbourg were still mighty sure of themselves and in the best of spirits. On the trip, Fleischmann proudly pointed out where the French-German border was—or should be, according to the German point of view. He took pride in the fact that the German language was spoken throughout a region which penetrated deep into France. But he was still angry at the local Nazi leader, the Gauleiter of Strasbourg, who had assured the people that there was no imminent danger of the city's being taken by the enemy.

"If I had only known," sighed Fleischmann. "All I had to do was cross the Rhine bridge."

Wouldn't we please let him go back to Germany so he could get even with that Dummkopf Gauleiter. Just think! A woman neighbor had gone across the bridge the day before to take some food to relatives and unsuspectingly had returned the next morning at ten. One hour later the Americans had taken the city. If she had known, she could have stayed on the other side of the Rhine. Oh, that scoundrel of a Gauleiter!

Fleischmann and his three companions were eventually taken to the United States. Although they resented their internment, they constantly praised the food and care, especially the medical care, which they were given. They probably missed it when they were returned to their beloved Germany some time in 1946.

It should be mentioned here that we found considerable information in Strasbourg about German war research in general, quite aside from nuclear research. The University had been groomed to become a model Nazi institution and most of the faculty had been selected because of their loyalty to the Party. Several were members of Himmler's SS, the Elite Guard. Thus the Professor of Anatomy, Hirt, was the official representative of the SS. It was he who furnished the other professors with concentration camp victims for their so-called scientific experiments.

In November, 1943, Professor Haagen, the virus expert, had written to Hirt:

73

15. 11. 1943.

Herrn Professor Dr. H i r t
Anatomisches Institut der
Reichsuniversität   G e h e i m
S t r a ß b u r g

    Am 13.11.43 wurden die mir von SS-Hauptamt zur Verfügung ge-
stellten Häftlinge einer Besichtigung auf ihre Eignung für die geplanten Fleckfieberschutzimpfungen unterworfen. Von den 100 Häftlingen, wel-
che in ihrem früheren Lager ausgewählt worden sind, sind auf dem Transport bereits 18 gestorben. Nur 12 Häftlinge befinden sich in einem Zu-
stand, der sie für die Versuche geeignet erscheinen läßt, vorausgesetzt,
daß sie zunächst in einen guten Kräftezustand versetzt werden. Hierfür
dürften etwa 2-3 Monate erforderlich sein. Die übrigen Häftlinge scheiden
infolge ihres Allgemeinzustandes überhaupt für den vorgesehenen Zweck
aus.

    Ich darf bemerken, daß die Untersuchungen die Prüfung eines
neuen Impfstoffes bezwecken. Derartige Versuche führen nur dann zu
einem brauchbaren Schluß, wenn sie mit einem normal ernährten und in
gutem allgemeinem Kräftezustand befindlichen gesunden Menschengut ange-
stellt werden, wie er dem Körperzustand der Soldaten entspricht. Mit
dem vorliegenden Häftlingsmaterial können daher brauchbare Ergebnisse
nicht erwartet werden, insbesondere auch, da ein großer Teil von ihnen
mit Leiden behaftet sein dürfte, die sie schon für die geplanten Versuche
unbrauchbar machen. Eine längere kräftige Ernährung und Ruhe würden
hier keine Änderung herbeiführen.

    Es wird daher gebeten, mir 100 Häftlinge im Alter zwischen
20-40 Jahren zu schicken, die gesund und körperlich so beschaffen sind,
daß sie vergleichbares Material liefern.

                         Heil Hitler!

               (Stabsarzt Prof. Dr. E. Haagen.)

Letter by Professor Haagen concerning the use of prisoners for medical experiments.

"Of the 100 prisoners you sent me, 18 died in transport. Only 12 are in a condition suitable for my experiments. I therefore request that you send me another 100 prisoners, between 20 and 40 years of age, who are healthy and in a physical condition comparable to soldiers. Heil Hitler!"

This was the same Haagen who had worked for several years before the war at the Rockefeller Institute in New York, at the same time doing a little extra-curricular work for the Nazi Bund. It was with distinct pleasure that the Alsos Mission traced him later in Germany and turned him over to the authorities.

In spite of the decisive finds at Strasbourg, neither the military nor our civilian colleagues, initially, were as convinced as we were that their fears of a German A-bomb attack had been unfounded. Was I absolutely sure that the papers we had found in Strasbourg were not purposely planted there to mislead us? Now that we knew where Heisenberg and his key men were working, wouldn't it be a good idea, and indeed imperative, to bomb the place? In fact, plans for the bombing were already under way, but we were able to prevent it by stressing the insignificance of the German project. Aerial photographs confirmed our statement that the project could not be of major importance. The main laboratory was housed in a wing of a small textile factory.

It was not until some time after Strasbourg that the full significance of our discovery dawned on me. "Isn't it wonderful that the Germans have no atom bomb?" I said to the Mysterious Major. "Now we won't have to use ours."

His answer, so utterly correct, took me by surprise. "Of course you understand, Sam," he said, "if we have such a weapon, we are going to use it."

That was early in 1945.

VII

We Meet Some German Colleagues

ABOUT the middle of March, 1945, Alsos, on the heels of our troops, crossed the Rhine from Ludwigshafen to Mannheim, and shortly thereafter entered the important university city of Heidelberg. Our first task was to occupy the laboratory of Professor Walther Bothe, Germany's foremost nuclear experimenter, in the physics laboratory of the Kaiser Wilhelm Institute for Medical Research. Twenty-four hours after our men had taken over, another contingent of U.S. troops arrived to take the Institute by surprise. It was they who were taken by surprise to find us already installed there.

As I made my way to the laboratory to see Bothe, I was aware that I did not quite know how to proceed. Here I was going to meet the first enemy scientist who knew me personally, a physicist who belonged to the inner circle of the German uranium project. It had not been too difficult to act authoritatively toward strangers, especially when I was backed up by a couple of real officers. But how could I be authoritative with Bothe, who was not only an old acquaintance and colleague, but certainly my superior as a physicist? How did one command an older and respected colleague to turn over his papers?

What I wanted most of all was information about their uranium research and no doubt my approach would have to depend on how the initial meeting with Bothe turned out. Would I have to have him locked up by our army men? Should he be interned and sent to the United States like the Strasbourg physicists?

Bothe greeted me warmly, and we shook hands, which was against the non-fraternizing rules.

"I am glad to have someone here to talk physics with," he said. "Some of your officers have asked me questions, but it is evident that they are no experts on these subjects. It is so much easier to talk with a fellow physicist."

He then proceeded to tell me about some of the research work done in his institute and he showed me reprints, proofs and manuscripts of all the war-time papers which were written under his direction. He took me around the laboratory, where we inspected the cyclotron, of which he was very proud. It was the only German cyclotron in operating condition, whereas the United States had about twenty of these important machines for nuclear research.

We talked in a friendly way for a long time.

I was surprised at the large amount of pure physics which was done during the war. There could hardly have been any time left over for war research. Finally I popped the question: "Tell me, Herr Kollege," I said, "how much did your laboratory contribute to war problems? It is obvious that not all your time was spent on the interesting work you have explained so far."

Professor Bothe became nervous. "We are still at war," he said. "It must be clear to you that I cannot tell anything which I promised to keep secret. If you were in my position you would not reveal secrets, either."

There was little I could say to that. I argued that the war in Europe was almost over anyway, and I dropped a few hints that I knew a lot about the uranium problem already. But the more I insisted, the more excited and angry Bothe grew. Plainly I was not getting anywhere with him. But then, previous experience had proved that one did not get very far merely by asking questions. The thing to do was to get hold of the documents. Strasbourg had yielded important information about the German uranium project, but we still lacked actual research reports with measurements and diagrams and results.

"I understand your reluctance to talk," I said to Bothe. "But I should appreciate it if you will show me whatever secret papers you may have. This, I believe, is a normal action in war."

Bothe shook his head. "I have no such papers,"

he said. "I have burnt all secret documents. I was ordered to do so."

Frankly, I did not believe him. He might have burnt official papers, but it was hard to believe that a physicist would burn the results of his painstaking research, no matter how many times it was stamped "secret."

But Bothe insisted that he spoke the truth. He had regretted the act, he explained, but strict orders had been given and that was that. Nor did a thorough investigation by our officers and counter-intelligence agents disprove what he had said. My suspicions were wrong. All evidence pointed to the fact that Bothe had told the truth.

He was a man of his word, and utterly trustworthy. He did not divulge any secrets until after VE-Day. Then, in July, he submitted a report giving a survey of his war research on the uranium problem. Of course, he knew that we had by that time rounded up everybody and everything having any connection with the German uranium work.

Bothe was a loyal German but never a Nazi. He lost his professorship at the University of Heidelberg when the Nazis transformed that school into a Nazi stronghold. Instead, he got a post at the Kaiser Wilhelm Institute for Medicine, also located in Heidelberg, where party policies did not have such a strong influence.

The director of that Medical Institute was the famous organic chemist, Richard Kuhn. When the chemists with our Mission, Professors Louis Fieser of Harvard and Carl Baumann of Wisconsin, met him, Kuhn was most co-operative. They had known him before and he welcomed them back in his laboratory. He told them that he had no connection with war work, but that it was all directed by the chemist Thiessen in Berlin. He had no secret reports and had merely worked on the chemistry of modern drugs. There was, he said, a very complete chemical library of the German Chemical Society hidden in a cave in central Germany and it would probably contain all war reports. This tip sent Louis Fieser on an adventurous cave hunt in the course of which he found a lot of things but no library. The library, as we learned later, was in a cave just outside Berlin in Russian-occupied territory. It contained no secret reports.

Richard Kuhn's record did not seem too clean to me. As president of the German Chemical Society he had followed the Nazi cult and rites quite faithfully. He had never failed to give the Hitler salute when starting his classes and to shout "Siegheil" like a true Nazi leader. Those who knew him well claimed that he did all this only to save German chemistry. He clung to his post as president of the chemical society, to avoid this important position's falling into the hands of a radical Nazi and a bad chemist. As he once stated to a colleague: "I am giving the Nazis no opportunity to turn me out." Kuhn may have succeeded in misleading the Nazis, but I felt he was being very clever with us, too. I could not believe that he was not familiar with important war work, although I had no time to look into this matter further. We knew, however, that he had been one of the administration bosses of German war chemistry, and later in Berlin Baumann discovered some valuable secret reports on applied chemistry which were no doubt familiar to Kuhn. Back in Heidelberg we had him picked up by one of our officers.

I showed him the secret reports, and reproached him for not having told us about them half a year earlier, when he knew all the time what we were after and when he acted as if he were co-operating with us. "Look," I said, "your name is all over these reports. You are one of the editors. Here you are mentioned again as attending a meeting. And here is a report on your introductory speech at the secret meeting on synthetic resins, where you impress the listeners with the necessity of winning the battle against the blockade and of winning the future for German chemistry after the war. You see, Herr Kuhn, our set of reports is not complete. You can no doubt get us the missing copies."

Kuhn was obviously in a spot and he promised to get a complete set right away. An officer took him home in a jeep and a short while later came back with the complete set of secret chemical periodicals. It was quite important. It contained articles on industrially valuable applications of chemistry, such as the production of plastics, asbestos, the use of coal tar, aluminum, cellulose, sulfur, etc. This rare set of documents is now in the possession of the American Chemical Society, and I am still sore for allowing myself to be fooled by Herr Kuhn. We could have got these documents in April instead of September.

The top physicist at the University of Heidelberg was a very old man named Lenard. He was one of the earliest and most rabid Nazis and the outstanding Nazi among the scientists. In fact, he had been a Nazi since 1918, long before anyone thought of Hitler. He agitated against the Weimar Republic and was once jailed for monarchistic speeches. His laboratory had long been forbidden territory for U.S. physicists and our Allies of the First World War. He once had done some outstanding work in physics for which he received the Nobel Prize in 1905. But he claimed that he and not Roentgen had discovered Xrays and he felt frustrated for not getting more recognition for his work. The increase of his political activities after World War I was paralleled by a decrease in the quality of his physics. He became more and more the political agitator and less and less the physicist. He finally received from the Nazis, including Hitler himself, the recognition he had longed for through so many years.

When the U.S. Army took Heidelberg, Lenard

fled the city. He probably feared that our men would shoot at sight such an important eighty-year-old Nazi. He hid somewhere in a nearby village and after a couple of weeks gave himself up to the U.S. military authorities. They asked the Alsos Mission for advice. Germany's greatest scientist, they informed us, had surrendered to the local officers in charge. What should they do with him? "Ignore him!" I said. This, for a Nazi, was a greater punishment than being tried in Nuremberg.

Lenard was responsible for Bothe's dismissal. Bothe's successor at the University of Heidelberg was a first-rate windbag and second-rate physicist named Wesch. "You are lucky that your first name is not George," one of his not so sympathetic colleagues once said to him. "Why?" asked Wesch. "Well, then you would be called G. Wesch." (—meaning "dither" in German.) This man was the top Nazi in Heidelberg and was promptly jailed by the U.S. Military Government officials.

Unlike Bothe, but like all Nazis, Wesch in jail offered his scientific services to the Allies at once. He wrote a long-winded pompous report about his secret war work including promises for future successes. In reality his work was below par and had been ignored by all other German workers in his field. In typical Nazi fashion his report was accompanied by a written "apology." Though he was a member of the SS, he claimed

84

not to have been a real Nazi. His defense was so typical of the Nazi mentality that we shall quote some of the passages directly. After promising to continue his research for the U.S. military he pointed out that a prison cell was not the best place to do this.

> *"Geheimrat Lenard,"* he wrote, "the great scientist, Nobel Prize winner, friend of Rutherford, honorary member of the Franklin Institute U.S.A.—whose scientific judgment is still fully reliable—regardless of his political opinion—designated me always as his *best* pupil.
>
> "To espouse my cause is a duty and an honor, as well as a service to science and technology, the more so while I can justify my political views at any time."

He then went on to enumerate his virtues as follows:

> "I. My *secret* contributions, etc.
> "1. Risked my life for the Allied Armies. Offered loyal co-operation with them.
> "2. At all times supported oppressed *Catholic* compatriots.
> "3. Helper, friend and protector of *Jewish* compatriots since two decades.
> "4. Frequent assistance to other compatriots in uncertain political situations and to foreigners.

"II. *Political*

"1. No political speeches or other agitator's activities.

"2. Gave limited information only about my scientific work on applied physics and radio.

"3. Action for expulsion from the SS was started against me because of point 1.

"4. My marriage was not recognized by the SS, a relative was removed to a concentration camp and died there."

It must have been some time in June, when I was back in Paris, that I got word from our Heidelberg outfit that Bothe had been jailed by the local counter-intelligence authorities. I ordered an immediate investigation. It turned out to be an unfounded suspicion in connection with activities of the "Wehrwulf," the fanatical Nazi underground. He was soon released on our recommendation. Bothe could readily forgive us for having jailed him for nothing; such things had happened to people under the Nazis very often. What made his short imprisonment almost unbearable, however, was that the counter-intelligence officials, having another professor on their hands at the same time and being short of space, had put Bothe in the same cell with his archenemy, Wesch.

86

VIII

The Breakthrough

IF WE had spent our time in Heidelberg reading records and correspondence, we would merely have found confirmation of what we knew already. What we needed now was not just general intelligence, but real technical data, and for this we had to go further into Germany.

We were waiting for the next move of our troops. We knew of a secret "pile" laboratory in a village in Thüringen, originally the laboratory sponsored by Army Ordnance. We also knew of the secret laboratory of the Heisenberg group in southern Germany, south of the city of Stuttgart. But we did not know which of these two places would fall into our hands first and it was impossible to find out in Heidelberg.

In Paris we learned that General Patton's Army was progressing at a terrific speed through central Germany and that the Thüringen village might fall any day. There was no time to lose. Phone communications were inadequate and considered unsafe for such top secret information. Fortunately, Dave Griggs, one of the civilian advisers to General Spaatz of the Air Forces, was a physicist friend of ours who was also a pilot.

Dave was perhaps the only civilian who was allowed to fly Army planes in Europe during the war. He had promised before to help me out whenever an emergency arose. This was such a moment. First he flew to Heidelberg, alerted Colonel Pash and party, and followed them to the village in Thüringen which the next day fell into our hands. He then flew back to Paris to pick me up and bring me to the secret laboratory.

This was the first German uranium pile laboratory we found. It was located in an old schoolhouse. The cellar of that place looked almost like a natural cave and seemed quite bombproof. It was there that our men found the few remaining physicists huddled together with their families.

The Germans were surprised to have visitors so soon. Colonel Pash had noticed immediately a large block of paraffin, an important aid in nuclear physics, which was lying outside the building.

"What is this black stuff here?" he asked, pointing at some objects that looked like briquettes.

Oh," they said, "that is nothing but coal."

Colonel Pash picked up one of the pieces and said, "It feels awfully heavy for coal."

Then they knew that our party was not there by pure chance, and they understood what we were after. The "briquettes" were obviously blocks of pressed uranium oxide.

Near the middle of the cellar the Germans had dug a deep pit. In that pit they had planned to build a structure of blocks of uranium oxide surrounded by heavy water. Next to it was a small pit with a heavy lead cover which once housed an amount of radium. When we arrived there was only a negligibly small sample left. In fact, the whole place looked rather empty and dismantled; just enough was left behind for us to reconstruct what had been going on.

The laboratories upstairs were still mostly intact. They merely reflected the pitiful smallness of the whole enterprise. There were some good instruments, mostly of a standard kind, but it was all on the scale of a rather poor university and not of a serious atomic energy project.

This place had been the laboratory of the nuclear scientists who at one time had been working under Professor Schumann of Army Ordnance. The director was the physicist Diebner. The chief co-ordinator of nuclear research, the physicist, Walther Gerlach of Munich, had had his administrative headquarters here. But these men had cleared out, and most of the equipment and material was gone. Only a couple of minor research workers were left behind. They told us that Gerlach had left long ago, but that Diebner and some of his key men and all the material for uranium research had been taken away only two days earlier.

Shortly before the fall of the village, they told

us, a number of Gestapo men with trucks and cars had come to the laboratory. They ordered Diebner to collect all he needed for continuation of his atom bomb research, to have it loaded on the trucks, and to follow them and take his key men along. The destination was kept secret, but it was believed they had been taken to the "Bavarian Redoubt" where the diehards among the Nazis were supposed to make a last stand. That was indeed what had happened, we learned later; Diebner was taken to Bavaria and ordered to complete his atom bomb research as soon as possible there. There were, of course, no laboratory facilities available; he was moved from place to place and finally gave up.

This incident shows that the SS was definitely interested in atomic research and that it had the full attention of the Nazi authorities. They must have expected a lot from it. It also refutes the statement often heard that the scientists had no support from the government. It was the scientists' lack of vision which prevented them from asking for more support, which they certainly would have gotten.

Among the documents we found, the most interesting were the reports on general progress as prepared by Gerlach. Again these confirmed our former conclusions; the German uranium project was only in a very initial stage. But the present information acted like the focusing of a picture on the screen. Before we studied these documents

(*Clockwise*) Goudsmit, Wardenberg, Welch, and Cecil reviewing documents found at Walter Gerlach's laboratory at Thüringen. It was their first uranium find. (*Courtesy of Brookhaven National Laboratory*)

we had a rather vague impression of the whole enterprise; now it all became sharp and well defined. Now we began to know names and dates and amounts of money spent on the project.

It took us a few days to complete our investigation. The village was gloomy. There was neither light nor heat, but it was, fortunately, not cold. We couldn't work evenings and to fight boredom we played cards by candlelight and a single lamp rigged up with batteries from the laboratory. The house we had requisitioned was gloomy too. It was well stocked with food and clothing but decorated in the favorite German way with all kinds of stuffed animals. There was one particularly dumb-looking bird, a goose of some kind, which practical jokers thought was just the right thing to put into somebody else's sleeping bag. Crawling into your bag in the dark and suddenly feeling this cold feathery stuffed corpse was not exactly a pleasant sensation. There were, too, numerous pictures of a large Saint Bernard around the house. The only trouble was the bedroom rug matched these pictures perfectly.

The village was full of displaced persons of all nationalities. They discovered a wine store-house, got drunk and started to riot. Our little outfit was asked to stop them, but we couldn't afford to get involved in this business; moreover, we were only about twenty men, including four civilians. We told the Germans that they had imported these people and that it was their problem therefore to get rid of them. Fortunately, regular occupation troops arrived two days later.

There was one phase of the uranium project which had remained obscure. Had the Germans ever tried to obtain pure U-235 by a centrifuge method? Strasbourg had revealed certain hints in that direction; we suspected that some work was being done, but the precise location of this work was a riddle. Now all this was cleared up. We discovered that the Germans had indeed started a centrifuge project on a small scale, originally at the University of Hamburg. Because of bombings they had moved at least three times, first to southern Germany, later back to the North. The latest location given was in the little town of Celle, north of Hannover.

It was the middle of April; north and central Germany were crumbling. An Alsos group under Major Fisher had been on the alert near Göttingen to enter that university town and the nearby village where we knew a principal office of the Reich's Research Council was located, the Osenberg office. We now had to dispatch a courier to inform them about the importance of the nearby town of Celle.

Two days later I flew up there myself. Arriving in Göttingen with the Mysterious Major we found that the various Alsos parties had been extremely successful. Everyone had gotten what he was after, information on aeronautical research, chemical research, and a lot more. The crowning achievement was the Alsos capture of Osenberg and his complete outfit plus a key man of Goering's aeronautical research setup, who happened to be visiting there.

Major Fisher with Doctors Walter Colby and Charles Smyth had moved on to Celle. The Mysterious Major and I followed them. Celle was right at the front in those days. Air battles and shelling were common occurrences. We had to drive through a wooded area which had been hit and was still burning. The road was one long line of British army trucks and tanks and armored cars bumper to bumper. Our little jeep managed to weave through occasionally, but not very fast. In the other direction moved thousands and thousands of displaced persons—French, Belgian, Dutch. They had taken anything on wheels— baby carriages, push carts, wheel-barrows—to carry their few belongings, usually including a mattress. Some were lucky and had stolen a horse-drawn cart or even a slow farm tractor which pulled three or four wooden farm wagons like a slow-moving train. There were men and women and a few children; they all wore rags and looked exhausted, but they sang and displayed their national flags proudly.

We found the secret centrifuge laboratory in Celle housed in a few rooms of a parachute-silk

factory. The beautiful white natural silk furnished war souvenirs for the families of most of
the Alsos members; we did not forget the poor
sergeants and officers and our secretary, who had
to stay behind in Paris.

The centrifuge laboratory was again on a very
small experimental scale. It had been planned to
build a whole battery of these centrifuges later.
The purpose was not to separate U-235 for a
bomb; the plans were certainly not adequate for
that. The Germans had never thought this practical on a short-time basis anyway. All they
wanted was to produce some uranium which was
slightly richer in U-235 than natural uranium.
If they would get this it would be easier to make
a uranium pile work with the limited supply of
heavy water they had. But the one centrifuge
that was in working order would have taken a
hundred years to produce any useful results.

The centrifuge laboratory was directed by Dr.
Will Groth. Almost twenty years earlier, when I
had a Rockefeller Fellowship for study in Germany, he had been my roommate. We had become very close and intimate friends. He was a
great admirer of Thomas Mann and that kept
him from falling into the trap of the Nazi doctrine. Being from Hamburg, he also had a more
international outlook than most Germans. But he
had done his duty; early in 1939, he and his colleague, Harteck, had written a letter to the Ger-

man military authorities about the possibilities of a uranium bomb.

Our meeting was sad; the root of our old friendship was still there, and we could not yet accept the idea that we were now facing each other as enemies. Both of us felt that we would rather have postponed our meeting until later, much later, or never.

I was happy to see that Groth had changed little in appearance as well as opinion. But our conversation did not flow easily. We exchanged some trivialities about our families, but both of us felt that there were so many more things to discuss of a deeper nature, for which this was neither the right moment nor the proper place. Groth took me through his laboratory and furnished a few superfluous explanations; I made the visit as brief as possible.

Our main objective had not yet been reached. The center of atom bomb research was not yet in our hands. We still had to contact the brains of the German uranium project, the physicist Heisenberg. We knew that he and his group were working in a make-shift laboratory in the village of Hechingen. Otto Hahn, the discoverer of atomic fission, and his men were in the nearby village of Tailfingen. Many other important research laboratories had also been evacuated to that region south of Stuttgart in the state of

Württemberg. A Swiss scientist, Dällenbach, had a separate laboratory there, "Research Center D," in which he intended to construct a cyclotron based on new principles. He had the full support and confidence of the Nazi research council; the Kaiser Wilhelm Institutes for Biology and Metallurgy were also near by.

We were waiting impatiently for this area to be occupied, when a new complication arose. This scientifically desirable territory was going to be taken by French troops. At the same time we had strict instructions to keep all atomic information closely guarded from all Allied personnel except the very small number of those who were already initiated. We were afraid that outsiders might stumble upon our valuable targets.

Colonel Pash got in touch with the military authorities and began extensive plans for a parachute attack on Hechingen. He planned to kidnap the physicists and carry away their principal documents. He began to train his inner circle of officers and men for this special mission. But I did not like the idea at all. We were absolutely certain now that the German atomic bomb project did not amount to anything. Though the parachute expedition did not seem to be especially hazardous, I considered the German project not worth even the sprained ankle of a single Allied soldier. With the help of the Mysterious Major I made a recommendation to this effect and the

96

parachute adventure was called off. I'm afraid Colonel Pash has not forgiven me to this day for having spoiled his interesting plan. It was to have been the crowning achievement of his career.

Still the problem existed; we had to get there first, just as on our other ventures, and without having to give explanations. In U.S. and British territories we had done fairly well. We had run into occasional trouble for not complying with established rules. We removed documents and people connected with A-bomb work, without properly filling out the necessary blanks at local headquarters and without waiting for approval. The Mission was officially "banned" from various areas a couple of times, but each time it was soon reinstated. We were very careful to comply with all rules and regulations when our activities covered subjects other than the atom bomb.

On previous operations, the Alsos military had attached itself to a so-called T-Force. These were the troops whose objective was the capture of persons and places that were not of tactical importance but of value to Intelligence, for example, Japanese consulates in Germany, high Nazi officials, and so forth.

We could not attach ourselves to a French T-force, if they had any, without lots of explanations. The decision was made higher up that we should form our own T-force this time. We were to borrow some troops from the U.S. T-force which was supposed to go into the Munich area and be entirely on our own.

Colonel Pash organized the campaign accordingly. Its official code name was "Operation Humbug." We had two armored cars, a caravan of jeeps and other vehicles. They proceeded south along the eastern bank of the Rhine and then penetrated eastward into the area where the secret laboratories were. There was no opposition; a few shots only were exchanged and some of the villages were summoned to surrender by telephone. It was probably the least dangerous of all Alsos advances.

On such expeditions our military would move in first and the civilians followed half a day or more later. At first I believed this to be a measure to protect our valuable lives. But I was mistaken; they did not care whether we were exposed to shelling or bombing. Their only concern was to make sure that the enemy would not capture us alive and make us reveal atomic secrets.

On this occasion Colonel Pash was accompanied by General Harrison, who was the Chief of Intelligence in that southern area. There was no French interference; the colonial French troops were more interested in pigs and chickens than in atomic scientists.

They finally reached the village of Hechingen and at once proceeded to Heisenberg's laboratory. It was not hard to find with the excellent

Goudsmit in Hechingen. (*Courtesy of Brookhaven National Laboratory*)

Colonel Pash giving orders to advance through Thanheim on the way to southern Germany. (*From left to right*) Pash, Sgt. Holt, and Cpl. Brown. (*Courtesy of Brookhaven National Laboratory*)

Colonel Pash (*far right*) and members of the Alsos Mission reviewing operations at Hechingen. Back row (*from left to right*): Maj. David C. Gattiker (Br), Lt. Cmdr. Eric Welch (Br), Dave Griggs, Capt Reginald C. Augustine, Wing Cmdr. Rupert G. Cecil (Br). Seated around table (*from left to right*): Lt. Col. Percy Rothwell (Br), Sir Charles Hambro (Br), Dr. Carl A. Bauman, Fred A. Wardenberg, Lt. Col. John Lansdale, Sir Michael Perrin (Br), James A. Lane, and Col. Pash. (*Courtesy of Brookhaven National Laboratory*)

A textile factory in Hechingen (south of Stuttgart in the Swabian Alps), the relocation site of the Kaiser Wilhelm Institute (1944). (From *Das Kaiser-Wilhelm-Institut für Physik, Geschichte eines Instituts* by Werner Heisenberg in *Jahrbuch Der Max-Planck-Gesellschaft zur Förderung der Wissenschaften E.V.*, 1971)

maps and aerial photographs which we had obtained. The Colonel and the General entered
Heisenberg's office; he was not there. But the first
thing they saw, to the consternation of the General, was a photograph of Heisenberg and myself
standing side by side. The photograph had been
taken when Heisenberg was my guest at Michigan in 1939. Egged on by Colonel Pash, the General almost began to believe that I could not be
trusted and that I had close contact with the enemy. At any rate I had to stand a lot of teasing
about this incident afterwards.

Colonel Pash established headquarters in
Hechingen and assumed the responsibilities of a
T-force commander. We had secret passes, as if
we were members of a club, "The Rendezvous
Club," Boris T. Pash, secretary.

Next a thorough investigation began. For this
final achievement of the Mission some of General
Groves' officers and some of our British colleagues
had come over specially.

Regretfully, I came too late to prevent our
guests from blowing up the cave in which the
uranium pile research had been located. It was
an utterly useless thing to do, but fortunately
harmless as all equipment was first removed. It
was the kind of error which caused our Army
to destroy the Japanese cyclotrons. We had also
discussed destroying the cyclotron in Heidelberg, but common sense prevailed and the recommendation of the scientists on the mission was

followed up. The trouble was that all plans were originally conceived under the supposition that German atomic bomb work was dangerously close to ours and should be dealt with accordingly. To this we must add the observation that the local military authorities cannot be expected to be competent in judging what is dangerous or important in this highly secret and specialized field. Thus a cyclotron is mistaken for a vital part of an atomic bomb, which it is not, and a harmless physicist is considered as a potential saboteur who can blow up New York City.

This point of view was difficult to alter, even after we knew how insignificant the German work was; in fact, no plans existed which covered such a possibility. We had orders to capture and intern all German atom bomb scientists. This was all right when Strasbourg fell, but I decided that it was utterly unnecessary to intern a man like Bothe. He did not know anything that might be of value to our own uranium project, nor did he know enough to give away secrets to the uninitiated. But now we were in territory to be occupied by the French. It was necessary to take along a few of the key men, if it were only to convince the colleagues back home that our deductions had been correct. There were still those who believed that the German colleagues could not have been so far behind. For that matter there still are.

IX

Operation Humbug

WHILE Fred and a few other Alsos members, assisted by British colleagues, studied our treasure of documents, I interrogated most of the physicists who were now in our custody. I still had to decide what to do with them, which of them could be safely left behind and which should be turned over to the military for internment. They were an interesting group of men and it was not easy to come to any one decision that would apply to all of them.

There was Otto Hahn, for instance, the discoverer of uranium fission, the basic process of breaking up the atomic nucleus which makes the bomb possible. It was obvious that I would be severely criticized if I did not turn him over to the military. Then there were Von Weizsäcker and Wirtz, two key members of the Heisenberg group. We had looked in vain for Von Weizsäcker in Strasbourg, where he had been a faculty member since the German occupation. Now we had caught up with him and his dossier was rather fascinating.

Von Weizsäcker's father, once Under-Secretary of State in Hitler's cabinet, had later become German Ambassador to the Vatican. In this capacity, he was occasionally able to protect Italian physicists who appealed to him as the friends of his son.

Von Weizsäcker, the son, was not a real Nazi, but like his father he was a real diplomat. He knew how to strike a compromise with the Nazis whenever it was expedient. He had the confidence of the Nazis, even of the Gestapo, and they came to him for information on physics and physicists. For example, the "cultural" department of the Gestapo, Section IIIc, sent him the following secret letter in November 1944:

> "As a notable pupil of Heisenberg I request that you send my office a short survey report on the present state of theoretical physics, how it is sponsored by government and political departments, and how to effect its possible application to so-called war research. Moreover we are in general also interested in your views on so-called 'German physics.'
>
> "I would be grateful if you would grant my wish as soon as possible and greet you with
>
> "Heil Hitler!
> (name unreadable)
> SS—Lt. Colonel"

During the war Von Weizsäcker was a kind of good-will ambassador for Nazi culture in foreign countries. He was sent to France, Spain, and Portugal, and after his return submitted reports on how he was received. For example, in his March, 1944, report on Spain and Portugal, he writes that he had close contact everywhere with the physicists, and he singles out a few for special mention. "Professor Palacios is scientifically the more significant, Professor Otero is conspicuous for his connections and sympathy for Germany. . . ." Von Weizsäcker further advocates closer cultural relations with Spain and Portugal and, though Germany would at first be the donor, "the cultural and political advantages which can still be obtained in these countries justifies a certain amount of effort." He was impressed by the "interest in German philosophy" and the "greater sympathy for Germany than in the other European countries, which I visited during the war." His reception in Paris, it seemed, was not exactly cordial.

The same month that Von Weizsäcker made this report, the German Foreign Office asked him to give a short radio talk, which would be beamed to the United States, about German science, because "enemy propaganda for several years has been spreading biased reports on the alleged decline of German science." He replied that his knowledge of English was insufficient for this assignment.

Von Weizsäcker played a prominent rôle in appeasing the pro-Nazi scientists who wished to ban modern physics and relativity from the curriculum because they were "non-Aryan." He always knew how to word things so that they were acceptable to both sides. During the war he seemed to have spent more time on such activities than on physics; his principal scientific work was an interesting new theory on the origin of the planetary system. Nevertheless he ranks definitely still among the outstanding theoretical physicists of our time.

In addition to Otto Hahn, Von Weizsäcker, and Wirtz, we also picked up two of the younger men, because they had done some novel research work on isotope separation. This puzzled Von Weizsäcker very much; he evidently thought that the young fellows were not important enough to be interned. "What kind of selection is this!" he complained.

The hardest decision to make concerned Von Laue. Unreliable rumors had originally designated him as a key man on the German uranium project, but all evidence indicated that he had nothing to do with it. During the entire Hitler regime, including the war years, Von Laue had openly opposed the Nazis in his actions and utterances. He never gave in and many of his friends feared that he was endangering his freedom and even his life. But his example showed clearly that with courage it was not necessary to bow to the

Nazi tyrants or lick their boots. When he visited Sweden during the war, he wrote letters to his friends in Allied countries describing the situation inside Germany. In a Stockholm lecture he mentioned Einstein's relativity theory, for which he was severely reprimanded on his return by our old friend SS Brigadeführer Ministerial Direktor Professor Doktor Rudolph Mentzel. Von Weizsäcker, the lover of compromise, advised him to answer his official critic with the statement that "The theory of relativity would have been developed without Einstein, but it did not develop without him." Von Laue refused to adopt this attitude; instead he published a paper on an application of the theory of relativity and wrote to Von Weizsäcker, "That is to be my answer."

He received the Nobel Prize in 1914 and is still one of the world's leading active theoretical physicists. Here was a man who had virtually been on our side throughout the war, who demanded the respect of his colleagues all over the world for his science as well as his personality. Such a man was indeed rare in Germany.

Nevertheless, I decided to have him interned along with the others. My thought was that he might be brought into contact with colleagues from our side in order to discuss the future of German physics. But although I made strong recommendations to that effect and the military treated him very well, my request was never quite fulfilled. I repeated my recommendations several times in official communications and, finally, directly after Hiroshima, I wrote as follows regarding Von Laue and Otto Hahn:

> "I strongly recommend that they be given an early opportunity to confer with a few prominent Allied colleagues on the general state of science in Germany before and during the war. They may have constructive suggestions about postwar matters. If any small-scale revival of German scientific education occurs, whether planned by the Allies or without our interference, it would be desirable to have these men in key positions."

It was a sad caravan of command cars and jeeps that departed at last for Heidelberg with our six "prisoners." Seeing them off, I could not help thinking of James Thurber's delightfully absurd cartoon titled "The Capture of Three Physics Professors." I felt a little like Thurber's fierce female pointing her gun at the sorry looking, droopy professors. At any rate, I was certainly glad that I had opposed Colonel Pash's plans for an airborne operation against the physicists. That might have out-Thurbered Thurber.

It was so obvious that the whole German uranium setup was on a ludicrously small scale. Here was the central group of laboratories, and all it amounted to was a little underground cave, a wing of a small textile factory, a few rooms in an old brewery. To be sure, the laboratories were

Capture of three physics professors

from James Thurber's *The War Between Men and Women*

Copyright by James Thurber. From *The New Yorker*. (*Men, Women and Dogs. Harcourt-Brace*)

well equipped, but compared to what we were doing in the United States it was still small-time stuff. Sometimes we wondered if our government had not spent more money on our intelligence mission than the Germans had spent on their whole project.

We were still lacking Heisenberg, our chief target among the German physicists. A few days before we took over Hechingen, he had left by bicycle to join his family in Bavaria and that territory was still in German hands. Before clearing out he had given strict instructions to conceal all pertinent materials and keep their whereabouts secret. But his precautions and instructions were in vain. We discovered the stuff that had been so carefully buried—all of it—two tons of uranium, two tons of heavy water, and ten tons of carbon, according to an AP press interview of Heisenberg.

The documents we had found gave us a lot of technical information we had long searched for. But I had the feeling that some documents were still missing. I could not believe that Heisenberg would have hidden only the uranium and not the important results of his researches. The more I considered the matter, the more I became convinced that we were lacking some important papers.

I finally got wind that the missing documents had been sealed in a large can and lowered in the outhouse of one of the physicists's homes. I called

Cubes of uranium dug up in field near cave-laboratory outside Hechingen. Here members of Alsos stack their first big haul on the grass. (*Courtesy of Brookhaven National Laboratory*)

Otto Hahn (*in coat and cap*) being taken away from laboratory in Tailfingen, 1945. (*Courtesy of Brookhaven National Laboratory*)

in a lieutenant and a GI and commanded a German to show them the way.

"Follow the instructions of this German," I said to the lieutenant. "This is a very important top secret assignment. Be sure your arms are in proper condition."

Not realizing what was in store for him, the officer was intrigued by his mission and expressed his appreciation of my confidence. However, he got his revenge; the last laugh was on me. He deposited his obnoxious find directly under the open window of the room where I was billeted. It was I who had to see that my arms were in order when I opened the container. As I had thought, it held the principal reports on the German uranium pile research.

Meanwhile our GI's had made a good find, too. They discovered an exquisite supply of wine belonging, presumably, to Von Weizsäcker's family. It was duly confiscated and, two weeks later in Heidelberg, consumed at our V-E Day celebration.

Early in June, the German physicists we had left behind in Hechingen tried to send a report about the events by messenger from the French zone through the U.S. zone into the British, where the directors of the Kaiser Wilhelm Society were located. In those days it was easier for Germans to travel than for Allied officials, who needed all sorts of passes and official orders to cross from one occupation zone into another. A copy of the report was of course intercepted by the Alsos Mission. I quote a few passages, because their unconscious humor is so typical of the German point of view. Note especially the order in which the losses to the Institute are enumerated.

"Hechingen was occupied by French and Moroccan troops on Sunday 22 April 1945 at 17 hours. . . . On the day of the occupation, four American tanks appeared about eight-thirty together with many trucks of the T-Force of the Sixth Army Group under the military command of Colonel Pash and the scientific direction of Professor Goudsmit. All laboratories . . . were occupied and searched and the members who were present were cross-examined. . . .

"In Hechingen the removal of apparatus was confined largely to that of the uranium problem, namely the two setups for separation of isotopes of Dr. Bagge and Dr. Korsching. In addition, souvenirs were taken, all the Institute's rubber stamps including the 'Minerva.' The loss of one thousand marks from the safe of the Institute was painful and the loss of five thousand marks from the safe of the firm Grotz, whose office was occupied by the Americans, was unpleasant.

"In addition to the apparatus and materials, Professor Von Laue, Professor Von Weizsäcker, Dr. Wirtz, Dr. Bagge, Dr. Korsching.

and Professor Hahn . . . had to accompany the Americans to an unknown destination for an unknown period."

After complaining about the many French commissions which visited the place later and about a German colleague who told them some of the uranium secrets, the report states, "Still we hope that an Institute of the rank of ours will be protected. It would bring honor to France and to the whole world."

Although the "Bavarian Redoubt," where the fanatical Nazis were supposed to make a last stand, was not yet in our hands, Colonel Pash decided he would get Heisenberg anyway. We knew that Heisenberg was in the little town of Urfeld, south of Munich, and the Colonel took six men and started out for the village through the mountains, avoiding the defended region around Munich. Entering Urfeld without mishap, he soon located his target.

Word got around quickly that an American colonel was in town and soon Pash and his men were visited by a couple of high SS officers. They wanted to surrender to the Americans, along with several hundred troops which were hidden somewhere in the surrounding mountains. They assumed of course that the Colonel would have a whole regiment under his command. Colonel Pash studied the map and told them to bring their men down to some specified places the next morning at a given hour. He then ordered his own officers to guard the various roads and effect the surrender. Unfortunately, one lieutenant, who was tired and not very attentive, suddenly blurted out, in the presence of the English speaking SS officers, "But Colonel, we are only seven altogether." An hour later the Alsos Mission had retreated in haste to a safer spot and did not return to Urfeld until a day or two later, when it was really taken by our Army.

A few days afterward Munich fell. Here Carl Baumann, accompanied by a few officers and men, located Walther Gerlach, who had been in charge of nuclear research and all physics research for the last year. He also discovered Diebner and the uranium, which the Gestapo had taken from the secret laboratory in Thüringen, and brought back more interesting documents.

I had just returned to Heidelberg when Heisenberg was brought in. I greeted my old friend and former colleague cordially. Purely on the impulse of the moment I said, "Wouldn't you want to come to America now and work with us?" But he was still too impressed by his own importance and that of his work, to which he ascribed his internment.

"No, I don't want to leave," he said. "Germany needs me."

He had said the same thing so often before, but now it was probably true. Provided, that was, his contact with Nazism had not changed him too much. He had already lost the confidence of several of his anti-Nazi colleagues.

"If American colleagues wish to learn about the uranium problem," he said, "I shall be glad to show them the results of our researches if they come to my laboratory."

It was sad and ironic listening to him say this, when I was aware how much more we knew about the problem than he did. But I could not tell him about the Allied progress, and so I did not contradict him. I merely thanked him for his offer and left him secure in the belief that his work was ahead of ours.

I had known Heisenberg for many years. He had visited me in Holland as far back as the summer of 1925. He had been in the United States several times and had spent summers with me at the University of Michigan. His last visit there, when he stayed at my home, was just before the war, in July 1939. He is still the greatest German theoretical physicist and among the greatest in the world. His contributions to modern physics rank with those of Einstein.

Heisenberg had openly fought the Nazi excesses. He had even succeeded, in 1937, in getting an article published in Hitler's newspaper, "Das Schwarze Korps," in which he defended Einstein's theory of relativity. He was severely attacked by the Nazis for this, particularly by an extreme Nazi physicist, Johannes Stark. In a leading article in the same paper Stark denounced Heisenberg and other theoretical physicists, calling them "White Jews." Stark, who, like Lenard, had been a Nazi long before Hitler came to power, received the Nobel Prize in 1919. His later work in physics, however, was so inferior that even the Nazis got wise to him. On the basis of his loyalty they made him president of their Bureau of Standards, but he did not last long in that job.

Heisenberg had many offers of positions in the United States. He was so convinced of his own important rôle in Germany, however, that he consistently refused to emigrate. He was always convinced that Germany needed great leadership and that he could be one of the great leaders. "One day," he said, "the Hitler regime will collapse and that is when people like myself will have to step in."

His extreme nationalism led him astray, however, during the war. He was so convinced of the greatness of Germany, that he considered the Nazis' efforts to make Germany powerful of more importance than their excesses. He still was stupidly optimistic in his belief that these excesses would eventually stop after Germany had won world domination. Near the end of the war, when visiting Switzerland and everything seemed definitely lost, he said, "How fine would it have been if we had won this war."

Although he fought courageously against Nazi excesses and especially Nazi stupidities, his motives were not as noble as one might have hoped from such a great man. He fought the Nazis not because they were bad, but because they were bad for Germany, or at least for German science. His principal concern was that Germany might lose its lead in science, especially physics. That is why he strenuously objected to the exile of German Jewish physicists.

His defense of the Theory of Relativity in Hitler's newspaper and the subsequent vile attack on him by Stark, caused him deep concern. This was not because of danger to his own person, but to the future of German physics. Progress in physics is impossible without the understanding and teaching of the Einstein theory, which is not a philosophical doctrine but an experimentally verified set of laws, like those of Newton, for example.

A family friendship with Himmler, together with the attack in "Das Schwarze Korps," gave Heisenberg an excuse to try to get in contact with the Gestapo chief. Himmler thought that Heisenberg merely wanted a better job, whereas what he wanted was to convince him of the necessity of having Einstein's theory taught to science students.

The result was that Heydrich, who later became the notorious hangman of Lidice, was or-

Der Reichsführer-SS   Berlin, den 21 Juli 1938

Tgb.Nr. AR/453

RF/Pt,

     1).   SS-Gruppenführer H e y d r i c h

        B e r l i n .

        Lieber H e y d r i c h !

           Den sehr sachlichen und guten Bericht über Prof. Werner Heisenberg, Leipzig, habe ich erhalten. Ich lege Ihnen einen sehr ordentlichen Brief des Prof. P r a n d t l, Göttingen, bei, dem ich sehr beipflichten muss. Ferner lege ich Ihnen meinen Brief an Heisenberg in Abschrift zur Kenntnisnahme bei.

           Ich bitte Sie, dem Reichsstudentenführer doch den Vorschlag von Dr. Prandtl, dass Heisenberg in der "Zeitschrift für die gesamte Naturwissenschaft" etwas erscheinen lassen kann, sehr nahe zu legen.

           Ich bitte Sie ferner, durch S i x den ganzen Fall sowohl beim Studentenbund als auch bei der Reichsstudentenführung zu klären, da ich ebenfalls glaube, dass Heisenberg anständig ist, und wir es uns nicht leisten können, diesen Mann, der verhältnismässig jung ist und Nachwuchs heranbringen kann, zu verlieren oder tot zu machen.

           Darüber hinaus hielte ich es für gut, wenn Six Prof. Heisenberg einmal mit Prof. Wüst zusammenbrächte. Leiten Sie doch diesen gesamten Schriftwechsel Wüst zu mit der Bitte, ihn nach Kenntnisnahme dem Pers Stab wieder zuzuschicken. Wüst soll dann versuchen, mit Heisenberg Fühlung aufzunehmen, da wir ihn für das Ahnenerbe, wenn es einmal eine totale Akademie werden soll, vielleicht brauchen können und den Mann als guten Wissenschaftler zu einer Zusammenarbeit mit unseren Leuten von der Welteislehre bringen können.

                Heil Hitler ! Ihr

                gez. H. Himmler.

Himmler's letter to hangman Heydrich about Heisenberg

116

dered to make a thorough investigation of Heisenberg's loyalties. On July 21, 1938, Himmler wrote:

> "Dear Heydrich,
> "I have received the good and very objective report on Professor Werner Heisenberg, Leipzig. I enclose herewith a very proper letter of Professor Prandtl, Göttingen, with which I agree. I also enclose a copy of my letter to Heisenberg for your information. . . . I believe that Heisenberg is a decent person and that we cannot afford to lose or to silence this man, who is still young and can still produce a rising generation in science."

And to Heisenberg he wrote on the same day:

> "Only today can I answer your letter of July 21, 1937, in which you direct yourself to me because of the article of Professor Stark in 'Das Schwarze Korps'.
> "Because you were recommended by my family I have had your case investigated with special care and precision.
> "I am glad that I can now inform you that I do not approve of the attack in 'Das Schwarze Korps' and that I have taken measures against any further attack against you.
> "I hope that I shall see you in Berlin in the fall, in November or December, so that

we may talk things over thoroughly man to man.

"With friendly greetings,
"Heil Hitler!
"Your,
"H. Himmler

"P.S. I consider it, however, best if in the future you make a distinction for your audience between the results of scientific research and the personal and political attitude of the scientists involved."

Early in the war, a colleague who complained to Heisenberg that the Nazis did not realize the true meaning of science, got the answer, "Maybe they do not realize it, but on the other hand they have the advantage of giving money if the plans one has are large enough. Perhaps in the near future I shall have an opportunity to see one of the leading figures in Germany. If I tell him that, for example, we need a new astronomical observatory and that it must be the biggest of its kind in Europe and should cost five million marks, he will tell Hitler and probably decide that fifty million is the right sum for this purpose and that the observatory should be completed in one year."

"Yes," the colleague added, "and when everything is ready he will appoint some inferior astronomer who is a good Nazi as its director and there you have your great observatory."

Der Reichsführer SS
Tgb.Nr. AH 4573
RF/Pt.

Berlin SW 11 , den 11. 7.1938
Prinz-Albrecht-Straße 8

Herrn Prof. H e i s e n b e r g
L e i p z i g O 27
Bozener Weg 14.

Sehr geehrter Herr Professor H e i s e n b e r g !

Ich komme erst heute dazu, Ihnen abschliessend
auf Ihren Brief vom 21.7.1937, in dem Sie sich we-
gen des Artikels im Schwarzen Korps von Prof. Stark
an mich wandten, zu antworten.

Ich habe, gerade weil Sie mir durch meine Fa-
milie empfohlen wurden, Ihren Fall besonders korrekt
und besonders scharf untersuchen lassen.

Ich freue mich, Ihnen heute mitteilen zu
können, dass ich den Angriff des Schwarzen Korps
durch seinen Artikel nicht billige, und dass ich unter-
bunden habe, dass ein weiterer Angriff gegen Sie er-
folgt.

Ich hoffe, dass ich Sie im Herbst -aller-
dings erst sehr spät, im November oder Dezember-
einmal bei mir in Berlin sehen kann, sodass wir uns
eingehend mündlich von Mann zu Mann aussprechen kön -
nen.

Mit freundlichem Gruss und

H e i l H i t l e r !

Ihr

H. Himmler

PS. Ich halte es allerdings für
richtig, wenn Sie in Zukunft
die Anerkennung wissenschaft-
licher Forschungsergebnisse
von der menschlichen und politi-
schen Haltung des Forschers klar vor
Ihren Hörern trennen.

Himmler's letter to Heisenberg

119

In Heidelberg, we had evacuated one of our beautiful villas to house our physicist guests, now nine in number, where we were holding them until we could arrange to take them to Paris and turn them over to the military authorities. They were under strict guard. We did not want the other physicists in Heidelberg to know of their presence. The only men available for guard duty at that time happened to be some excellent American Negro troops. This did not please our guests at all; they had perhaps lived too long under the myth of Aryanism. Aside from this, they all seemed to be in excellent spirits. They considered themselves very important personages who had something to offer us that we wanted very badly. It was only later, on August 6, that they were rudely awakened out of this pleasant dream when the radio brought them the news of Hiroshima.

Walther Gerlach looked haggard, so much more worn than when I had last seen him in England in the summer of 1938. We were guests at the meeting of the British Association for the Advancement of Science. I had known him well, for many years, but in England he acted evasive. I invited him to visit me in Holland on his way back. He made a vague promise, implying that such a visit required special permission from the Nazi government. Altogether he was noncommittal and acted scared when I tried to talk to him about the political situation in Germany.

Now he was far more cordial; he was again the Gerlach I had known before the Hitler regime. It is true that he used to act like an absolute dictator in his laboratory and that students and assistants feared him, but they loved it. He was quite talkative and gave us a rather objective account of science under the Nazis. He was an excellent experimenter, lecturer, and leader. During the short year that he had replaced a loyal Nazi named Abraham Esau as chief of all physics research he had done his best to save what he could. He was not a Nazi, but on many occasions, we learned, his judgment had failed him and in order to avoid trouble he had played into the hands of the Nazis, especially in the early days of their regime. He himself had been denounced too, for personal reasons, and was tired of interference. His only wish was to save and to promote German physics without the help or obstruction of the Nazis. His principal contacts were with officials in the Speer Ministry of war production, who backed his efforts. But it was all too late. The war was over before Gerlach's influence took hold. He tried to be objective and recognized some of the merits of the research undertaken by Diebner, whose Army Ordnance group had competed in uranium experimentation with Heisenberg. But his support of Diebner met with the disapproval of the smug Heisenberg clique.

We could get nothing out of Diebner. He was as sullen as a real prisoner. He must have felt like an outcast, living in the same house with members of the Heisenberg clique. Their conversations with him were limited to monosyllables.

One key man was still unaccounted for, the physical chemist Harteck. The research reports, which we had begun to study, showed that he was perhaps one of the most active and far-sighted members of the Uranium Club. His specialty was the separation of the uranium isotopes and the production of heavy water. He had long since realized that these projects should be put on an industrial basis as they had outgrown the academic laboratory stage. Later Major Russell Fisher located him in Hamburg and brought him to the Alsos headquarters in Paris. This "kidnaping" without the proper formalities, which we were compelled to do because of atom bomb secrecy, brought down on us the severe criticism of the military authorities in that zone; somehow they had learned of the Alsos Mission's illegal activities in their territory but apparently had not guessed just what we were there for.

Harteck was brought to my office. I had not met him before. At first he was rather reluctant to talk. He was not aware of how much we already knew about him and his work.

"Tell me something about your efforts to make volatile uranium compounds," I said, knowing that he had undertaken this research in the hope of finding some substance suitable for isotope separation by gas diffusion methods.

Harteck admitted having directed such work. "It was not successful," he said. "The nearest we came was with some complicated organic compounds containing a uranium atom surrounded by atoms of hydrogen and carbon. Very complex molecules indeed, hard to describe," he said. "I'll try to show you," he continued. Pointing at the paperweight cube on my desk he said: "Now let us assume that that represents uranium." He moved to pick it up and its weight took him by surprise. "But this *is* uranium!" he cried.

That incident ended his hesitancy. He recognized the uranium cube as coming from the German pile experiments, and he proceeded to give a complete account of what he knew.

This was the end of our roundup of the German nuclear physicists. We had found all the key men on the project and all documents and materials. We had interned fourteen of them, of whom four were already in the United States; the remaining ten were held by the military temporarily at Versailles in "Dustbin," the internment center for important civilians. There were only a few nuclear physicists left in Berlin or in the Russian zone of occupation, but we had talked to a couple of them in Leipzig, when that city was open to U.S. personnel.

It was not until late in July that a small Alsos group was allowed to enter Berlin. As we expected, we found no new information but what we learned was very satisfying. It was like the last pieces of a jigsaw puzzle; the pieces of haphazard information we gathered completed the picture, plugged up a few minor holes, but the pattern remained the same.

We found, for instance, the chief chemist of the Auer Chemical Company, for whom we had been looking ever since we had entered Belgium. But he could tell us nothing we did not already know, nor could the few industrial physicists who still remained in Berlin. The Gestapo scientists had all cleared out before our arrival, some of them leaving sufficient clues in their deserted homes for us to track them down later. And then there was Rudolph Mentzel's secretary. She was living alone in the porter's lodge of a large burnt-out and bombed-out villa, which had once housed Himmler's own scientific academy, the "Ahnenerbe" or Academy of Ancestral Heritage. The basement, the only place not completely ruined, still contained remnants of weird Teutonic symbols and rites, strange dummies which at first sight looked like bodies of victims, a corner with a pit of ashes in which I found the skull of an infant.

The Egyptian Museum was hardly the place to look for uranium clues, but I had visited the place in the old days when it was one of the finest museums of its kind and I was curious to see what had become of it. The place was an utter ruin. About the only thing left in it was a bomb-dam-

Goudsmit and prewar friends at the University of Michigan. (*From left to right*): Goudsmit, Dean Yokum, Heisenberg, Fermi, Dean Kraus. (*Courtesy of AIP Niels Bohr Library*)

A small Alsos group examines some "junk" left by Germans as they evacuated the Kaiser Wilhelm Institute of Physics in Berlin. (*Courtesy of Brookhaven National Laboratory*)

aged, sadly dilapidated looking mummy of the Ptolemaic era. The poor old watchman who had been with the museum in its better days was so impressed that anyone remembered the place, he insisted that I take the mummy with his compliments. For a moment I was tempted to do so and send it back to Washington with instructions to "test for radioactivity." But it was too big for our jeep and I had to be content with a few fragments of the painted mummy wrapping.

Our chief visit was, of course, to the now empty Kaiser Wilhelm Institute for Physics, where the uranium research had started in 1939. It was one of the few buildings still wholly intact. A big sign on the door informed us that the Director of Intelligence of the U.S. Control Council had set up headquarters there. The place seemed empty. We went in and found one room furnished with two desks and one officer. We had come across him before. He did not understand our interest in this building.

"It's all empty," he said. "Everything, even switches and wiring, has been removed by the Russians. We found some junk which we dumped in the back yard. The sub-basement looks queer. It seems to have been a swimming pool. Go around and take a look."

We inspected the place thoroughly. The backyard "junk" contained various pieces of equipment for nuclear physics as well as blocks of pressed uranium oxide. There were also some notebooks indicating the type of research that had been going on. The sub-basement was the bomb proof "bunker" laboratory of which the Germans were so proud. It looked as if it had once been excellently equipped. The "swimming pool" was the pit in which the pile had been con-

The underground uranium laboratory of the Kaiser Wilhelm Institute for Physics in Berlin, as found by the Alsos Mission.

structed. Metal containers and frames for the arrangement of the uranium cubes were still standing near by.

I remembered the primitive setup with which Enrico Fermi had started in a basement room at Columbia University. By contrast, this Berlin laboratory, even empty, gave an impression of high-grade achievement. I stood there alone for a little while and in the dim light thanked God for the great privilege of being permitted to see with my own eyes, and in a language I could understand, the physicist's symbol of the defeat of Nazism.

X

Hiroshima and the German Scientists

ONE day early in August, 1945, while I was looking through the ruins of Himmler's headquarters in Berlin for any clues I might find of his interest in scientific research, I heard someone calling me: "Hey Sam! Sam, where are you?"

It was one of our Alsos officers, and he was quite breathless. He had been looking all over Berlin for me, he said, and had finally spotted my jeep with the Alsos emblem in the shambles that had once been Prinz Albrecht Street. He took a hurried glance at his wristwatch.

"You've got just fifteen minutes to catch a plane for Frankfurt," he said.

"Why should I go to Frankfurt?" I said. "I've just asked for more people to come here and help me. There's still plenty to do in Berlin."

"Orders," he said. "There's a special plane waiting."

"Can't I even get my toothbrush and pyjamas?"

"Absolutely no time. We've already lost more than an hour looking for you."

It sounded important. Dirty as I was—and one

felt really dirty rummaging through Himmler's leftovers—I climbed into the jeep and we raced through Berlin with sirens screaming to the Tempelhof airdrome, right through the gate without formalities and up to the plane. The propellers were already turning and in the open door stood a GI who had been looking for us anxiously. I climbed from the jeep into the plane, the door slammed shut, the motors roared and we were off. It was just like the movies.

I had no idea why I was wanted in Frankfurt where our Intelligence headquarters was located, or what all the rush was about. The crew of the plane were all strangers to me. There was no use questioning them; very likely they didn't know any more than I did. There will probably be someone to meet me at the airport, I thought.

But when I landed at Frankfurt a few hours later there was no military band to greet me, not even one of our Alsos jeeps. I got on an Army bus which took me to headquarters.

"Well, here I am," I said. "What's up?"

Everybody was wonderfully evasive. "Those fellows you sent to Munich," they said. "They're expected back and we thought you might want to give them instructions for their trip to Vienna. There are also a couple of officers coming in from London and maybe from Washington."

All of which did not add up to any satisfactory explanation for my melodramatic exit from Berlin. Even Colonel Pash put on a poker face when I questioned him. I was so mad I wanted to take the next plane back to my ruins.

However, there was one bit of consolation. Carl Baumann, Dick Beth and our secretary Mary Bohan arrived from Paris on their way to join me in Berlin, and they brought me two whole weeks of clean laundry. Food supplies were no problem for the Alsos Mission; there was always enough to eat. But because of our frequent movements from one city to another, we often suffered from a lack of clean underwear. At one time, in Göttingen, things got so bad that Allan Bates, our metallurgist from Westinghouse, searched the German house we were occupying. All he could find was a woman's long pink undies, which he was compelled to wear for several days—after modifying it slightly for male use.

After dinner that night, in the course of a long bull session at the billet of one of our civilian scientists, I learned for the first time that the A-bomb test in New Mexico had been a success. I learned, too, that the officers of our military contingent had also known about the successful test, but had told me nothing about it. Of course I had known for some time, through unofficial channels, that a test was planned and even the estimated date. But I felt rather piqued that I had been kept in the dark about the actual occurrence, a little as if I had been kept out of things deliberately. And of course it was deliberate, so that I could not give away anything when I was talking to the German physicists.

We went home a little before midnight. I was accompanying Mary Bohan to the WAC hotel where she was billeted and had stopped in the lobby to say good night. The sergeant at the desk had his radio blaring. Midnight struck. And then suddenly the loudspeaker burst out with the news of the atom bomb and Hiroshima.

I was amazed at the detailed information given to the world at large about what had been the top secret for so long. Miss Bohan was even more amazed. The secrecy had been so complete that even she, the Alsos secretary, had had only the vaguest idea of what we were after. It was only now that she began to realize what kind of outfit she had been working for. Past events, which she had not understood at the time, suddenly began to make sense for her. She overwhelmed me with questions, some of which I could not answer for security reasons and some because I just didn't know.

At any rate, now I had the answer for my sudden forced evacuation from Berlin. My military guardian angels in Washington were afraid that the news of the bomb might put our Russian colleagues on my tail. There were no barriers in Berlin between the Russian zone and the American. They feared that the Russians might kidnap me and force me to tell the atomic secrets. This was very flattering but rather foolish. Not even the Russians could force me to tell what I didn't know. And it would take months of bureaucratic wrangling before they knew who I was and where I was and how to approach me.

I returned promptly to Berlin and the only Russian who contacted me was a soldier who wanted to buy my watch. As luck would have it, my watch had stopped that very morning, and so my dream of making an easy two hundred and fifty dollars came to nothing.

Where I really would like to have been, that night a stunned world first learned of the atom bomb, was in England among the interned German physicists. After we turned them over to the military in Paris, they had moved from place to place until finally they were housed in an estate about fifty miles from London. There they were very well taken care of. They had good food, newspapers, radio, a piano, and tennis courts; they had even received new clothing. But because of the secrecy that surrounded everything connected with the atom bomb, their whereabouts was kept secret, too. Colleagues would ask me where Heisenberg was hidden, or Hahn, but I could not tell them because I had only a vague notion of their whereabouts myself, although unofficially I was kept informed about their general status and reactions. Their only complaint was that the special restrictions imposed upon them made communication with their families far more difficult than for ordinary prisoners of war.

All this hush-hush was necessitated by our original assumption that the Germans had the atom bomb or must be close to its secret. As it turned out, they knew practically nothing of significance. But by tracking them down and making such a thorough investigation, we might have shown our hand. Actually, the German scientists were so sure of their own superiority it never occurred to them that we might have succeeded where they had failed. But our military security experts could not be sure of this; they could not be sure that if these men were set free, the supposition that we might have a gigantic uranium project would become a matter of common knowledge everywhere. The risk was too great; the only thing to do was to segregate the men and keep their colleagues and the rest of the world guessing.

Just why these top German physicists were interned in England, I never understood. If they had been in the United States, I might have had more luck with my recommendation that at least Von Laue and Hahn be consulted about the future of German physics. Perhaps our military experts did not know what to do with these scientists after we had found them, and felt quite grateful when the British offered to take them over. As a result, the majority of the best German scientific brains is now in the British zone of occupation. Several excellent scientists are in the French zone. That there are still a few good physicists in the American zone is due to the fact that the Alsos Mission did not intern them.

It was on August 6, 1945, at dinner time, that the interned German physicists first heard the news of Hiroshima. Their initial reaction was one of utter incredulity. Impossible, they said. After all, they themselves had been working on the uranium problem for several years and they had proved that an atomic bomb was too difficult to achieve in such short order. Then how could the Americans do it? It was preposterous.

"It can't be an atomic bomb," one of their number said. "It's probably propaganda, just as it was in Germany. They may have some new explosive or an extra large bomb they call 'atomic,' but it's certainly not what we would have called an atomic bomb. It has nothing whatever to do with the uranium problem."

That being settled, the German scientists were able to finish their dinner in peace and even partially digest it. But at nine o'clock came the detailed news broadcast, the same one, very likely, that I heard at Frankfurt.

The impact on the ten scientists was shattering. Their whole world collapsed. At one stroke, all their self-confidence was gone and the belief in their own scientific superiority gave way to an intense feeling of despair and futility. Then all their work of the last six years had been wasted;

their hope for a bright future for German science was only so much illusion.

Only one of the group was not affected, at least not personally, and that was Von Laue. He had been merely an onlooker, never a participant in the physicists' dream of power, the atomic bomb. For the atomic bomb meant power not only for the country that could solve this problem, but power as well for the physicists and their science, which would now be recognized as indispensable. Von Laue was probably the only man among the ten interned on that English estate who realized thoroughly the world-shaking effects this Hiroshima bomb would produce everywhere, in ever-widening circles.

Von Laue, at any rate, took the news calmly. Not so the others. There were bitter words as the question arose why they, the superior Germans, had not succeeded. Some of the younger men angrily reproached their elders for not having had more vision, for having failed Germany in her hour of need.

Gerlach was the most violently upset of all. He acted like a defeated general. He, the "Reichsmarshal" for nuclear physics, had not succeeded in his assignment. He took the remarks of the younger men as a personal criticism and was profoundly depressed for several days. His colleagues and friends had to reason with him and comfort him to bring him back to his senses.

The rest got over their hysteria. They spent

hours discussing the science of the bomb and tried to figure out its mechanism. But the radio, for all its details, had not given enough, and the German scientists still believed that what we had dropped on Hiroshima was a complete uranium pile. No wonder they were bewildered. For us, or anyone, to have dropped a complete uranium pile would indeed have been a super achievement of modern engineering; as yet, no plane has ever been built that could do that. Even if there were such a plane, a uranium pile could never be a bomb. It could only be a fizz. But as yet, these Germans experts failed to realize this basic fact.

Failing to understand it, they began to carp at details. Why wasn't Otto Hahn, the discoverer of fission, given credit on the radio and in the newspapers? Why did the papers erroneously state that Lise Meitner had discovered the splitting of uranium and stress that she was Jewish? But especially, and above all, why didn't Germany succeed in making the bomb? They looked for all sorts of excuses and rationalizations. The Nazi government, they said, would never have backed the scientists the way the Allies did. Apparently, they had conveniently forgotten the active interest the Gestapo and other government agencies had taken in their uranium work. They forgot that they themselves had not been very convinced of their own chances of success; had the conviction been there, they would probably have had all the government support they could use.

But even if we had had more government support, some of them argued, we would never have given Hitler such a devastating weapon. This may have been true of Von Laue and Hahn, but it is extremely doubtful if it would apply to the rest of them. At any rate, intentions in the long run are fruitless; there are always enough co-workers who fail to see the consequences of their handiwork as clearly as the few great minds with vision and ethical scruples. After all, the atom bomb would have made Germany strong, and even if a Von Laue gave his life to prevent Hitler's having the bomb, the sacrifice would have been in vain.

But why did the newspapers and radio make such a point about the Allied destruction of the heavy water plant in Norway? Didn't they understand that heavy water had nothing to do with making a bomb? It could be used in making a uranium engine, certainly, but not a bomb. And what was this about plutonium? What was plutonium? The German physicists were more confused than ever. It must be those stupid newspaper reporters and radio announcers, they said; there is no such substance as plutonium. Maybe what they meant was polonium or protactinium, which had long been known as radioactive elements. But what part could they play in the making of an atom bomb?

It was not until more than a full day after the

first announcement of Hiroshima that Heisenberg began to understand how he and his colleagues had completely missed the basic principle of the atom bomb. It was only then that he finally came to understand that we had used the uranium pile merely to produce material—plutonium—and out of this new substance had made the bomb. The pile itself was never intended to be a bomb.

Heisenberg called his colleagues together and explained to them what it was all about. They were amazed, and crestfallen. It was all so simple. How could they ever have missed it? And how could they ever survive such a blow to German scientific prestige?

It was more than an academic matter. Their very lives might be in danger, if they were sent back to Germany now and fanatical nationalists placed the blame on them. Their anxiety to return home cooled perceptibly and they suddenly found English hospitality very comforting indeed.

It was then that some of the younger men hit upon a brilliant rationalization of their failure. They would turn that very failure to their advantage by denying that they had ever tried to make an atomic explosive. Only a few of the secret communications had ever even mentioned the word. They would stress the fact that they had been working only on a uranium machine, and forget that they had thought this would lead directly to the bomb. They would tell the world that German science never, never would have consented to work on such a horrid thing as the atom bomb.

This, then, was to be the new theme song of German science: "Germany worked on the uranium problem for peaceful uses only; the Allies, for purposes of destruction." It was therefore no surprise, at least not to the members of the Alsos Mission, when, almost two years after Hiroshima, Heisenberg spoke to an Associated Press reporter about "Germany's uranium pile, which I was building up to create energy for machines and not for bombs. . . . As the world now knows, the explosive, plutonium, is produced in such a uranium pile."

Heisenberg's statement is a beautiful example of how to use half-truths. It is true that the German scientists were working on a uranium machine and not the bomb, *but it is true only because they failed to understand the difference between the machine and the bomb.* The bomb is what they were after. And what the whole world knows now about plutonium, *the German scientists did not know—until they were told about it after Hiroshima.*

XI

The Misorganization of German Science

THE myth of the German genius for organization dies hard. Those who still believe that in war science, at any rate, the Germans had an admirable capacity for level-headed, far-sighted, efficient planning and execution are in for a grave disappointment. True, they did have an abundance of card indexes, blanks, charts and classifications, of bureaus and departments and titles, but when it actually came to getting things done they did not exactly belong at the head of the class.

In the United States, all wartime research was organized in the O.S.R.D.—Office of Scientific Research and Development—under Dr. Vannevar Bush. This organization, comprising practically all research engineers and scientists in the country, was made up of a number of divisions according to subjects—radar, rockets, or whatever the subject might be. Within each such division there was complete co-operation between the research workers of the armed forces, industry, and the academic laboratories. Moreover, research scientists were allowed to observe their products in action at the battle fronts. Special scientific groups worked with the Army and Navy to advise on the proper use of new equipment and evaluate its success. The only criticism occasionally voiced against the O.S.R.D. was the lack of sufficient exchange of ideas between the different subject divisions. On the whole, it was a gigantic setup that worked with superb smoothness and efficiency. It was this organization's Office of Field Service that recruited the civilian scientists for the Alsos Mission.

In Germany there was no over-all organization comparable to what we had in the United States. Nor had the research scientists any liaison with the armies in the field. It is a rewarding lesson to go back and study the organization—or misorganization—of German science, as it was revealed to us in the documents uncovered by Alsos.

Before the war the Nazis had voiced their prejudices against science and scientists loud enough for the whole world to hear. The cold logic of science did not fit very well into the mystic cult of blood and soil or the "certainties" of intuition. When the war started, the Nazis were confident they could win it in short order, so there was no need for scientific war research. All scientists were drafted indiscriminately; some of the best of the younger men were killed in action. Only a few of the most important were released after a short period of military service. The only people who escaped the draft entirely were the elect on Hitler's "Führer List," which contained about a thousand names of artists, dancers, movie stars and, probably, astrologers—but no scientists.

At the outset, there were several independent research organizations, with no practical co-ordination between them. The Germans, moreover, made a sharp and fatal distinction between "research" and "development." The latter was primarily the concern of industry and was, therefore, under Adolf Speer, Minister of War Production. The Army scientists knew almost nothing of what the Navy scientists were doing, and the Industrial engineers were never allowed to see their products in actual use.

Army research was conducted by the Ordnance Department headed by the mediocre physicist, Erich Schumann. Professor Schumann's right hand man was Diebner. These were the two men who, early in 1939, started secret atom bomb research for the German army, without the knowledge of reputable German nuclear scientists. It was they who visited Joliot-Curie immediately after the occupation of Paris with the intention of confiscating and removing his equipment for nuclear research. They were reported, also, to have visited the famous Russian nuclear physics institute at Kharkov, when that city fell into German hands.

Schumann was actually professor of military

physics at the University of Berlin, although his few publications deal only with the vibrations of piano strings—an interest derived, presumably, from the fact that he was a descendant of the composer, Schumann. His colleagues, somewhat contemptuously, referred to him as Professor of Military Music.

Incidentally, Schumann was director of the Second Physics Institute of Berlin University. Why second? To the uninitiated, this must seem as much a mystery as Shakespeare's "second best bed." But it is not really a mystery. In Germany everybody had to be a director of something; every professor felt that he must be a director of an institute. At the larger universities, where more than one physics professor was needed, the laboratory was divided into sections, which they called institutes, and thus each professor became the director of an institute. Occasionally the various institutes had separate buildings. More often they were merely the separate floors of the same physics laboratory, and the director of the "Third Physics Institute" was merely the boss of the basement or the attic. Of course each director was jealous of the other, allowed no trespass on his domain, and professed to know nothing of what the other fellow was doing. In Schumann's case, the work had been shrouded in secrecy even before the war, and so no one knew quite what was going on in the Second Institute, although the first rate physicists knew, from the type of personnel he was using, it could not be very important or successful.

Secret Gestapo reports—there was apparently a Gestapo spy present in any and every group—said that Schumann was unfit for the responsible position he held, that he had neither the knowledge nor the personality for it. But this opinion did not seem to have any effect on Schumann's standing with the Army. During the war, he rose to the post of personal adviser on scientific research to the Chief of Staff, General Keitel. In this capacity, he was nominally in charge of all research done by the armed forces and represented Keitel on important scientific committees.

If Schumann was the first to begin work on the uranium bomb problem, he was also the first to abandon it. Early in 1939 he had put his man Diebner to work on this project at the ordnance proving grounds at Kummersdorf, near Berlin. There, in a small underground shelter, they tried their hand at uranium pile experiments, more or less in competition with the academic scientists who had started similar work in Berlin and knew nothing about the Army research.

But the uranium problem is rather more difficult than the mysteries of piano strings and Schumann became impatient. By the end of 1942 he had lost interest in the project; he turned Diebner, personnel, equipment and material over to the civilian research organization, the Reich's Research Council, which had just been placed under Goering. He did not, however, turn over the two million marks his research group had been granted by the Army.

Schumann next devoted his talents to bacterial warfare. It is probable that in this field his competence was even less than in physics and its wartime applications. But he liked to be involved in things that looked important and his name shows up on many rosters of research committees.

When Berlin fell, Schumann fled to Bavaria. The Alsos Mission followed his trail for a short while, mainly out of curiosity, but we soon gave up. He was so obviously unimportant.

There was one branch of Army Ordnance which worked successfully. That was the V-2 group at Peenemünde on the Baltic Sea. A Gestapo report shows that originally this research group suffered from violent disagreements between civilian experts and Army officers. Dated June 29, 1943, this report gives us a fine behind-the-scenes view of a situation that occurred fairly frequently, and not always on the German side of the fence.

"In January, 1943, there was a serious disagreement about the continuation of the Peenemünde Project. Involved were, among others, General Keitel (chief of staff) and General Dornberger (in command at Peenemünde).

"The Führer called in all participants af-

145

ter each one was first obliged to state his point of view in writing. All military men were shown out within one or two minutes, because they were unable to answer the decisive questions of the Führer. Dr. Von Braun (technical chief at Peenemünde) was the only one who talked for 30 minutes and was able to answer the precise questions of the Führer tersely and clearly. The Führer decided, therefore, according to the proposals of Dr. Von Braun.

"It is said, in addition, that the Führer has not yet a clear enough knowledge of the daily little wars between officers and engineers in Peenemünde. . . ."

The report goes on to complain about the low priority which this important project had, and the consequent delays in the procurement of materials. (Directly after the report it was given the highest possible priority by Speer, the Minister of War Production.) It also mentions the names of two good physicists at Peenemünde, a Dr. Schilling and a Dr. Elvers, and adds this revealing comment: "The latter deserves a responsible position but is merely an anti-aircraft sergeant and thus cannot be placed high in this military establishment."

The Peenemünde Project did succeed; at least, the V-2's were used against London late in 1944. From a technical and scientific point of view these Buck Rogers inventions were really a marvel, but it is questionable if they could have altered the outcome of the war, even if they had been used earlier. The amount of explosive each V-2 carried was comparatively small; all the V-2's together would not have been equivalent to one major Allied air raid. Moreover, the V-2 was effective only against a target like London. It was the first step in a development which later may become of major importance in warfare, but the Germans used it prematurely. The only real advantage they had was that they could not be shot down like planes and did not involve the loss of highly trained personnel, like bomber crews. So maybe the generals Hitler kicked out after two minutes were right after all. They probably knew of better ways to use the men and brain-power wasted on this fantastic "revenge weapon."

In contrast to the Army research it must be said that the German Air Force research under Goering was excellently organized and highly successful. There are several reasons for this, the principal being that the Air Force was interested in results and not in politics. The committee in charge of Air Force research was selected on the basis of ability. With one lukewarm exception, the men were not Nazi party members. The men in charge were also aware of the great shortcomings of the other German research organizations and avoided as much as possible any contact with them. Thus, research on practically all subjects of interest to the Air Force was done by their own organization; they had divisions on radio, radar, aircraft engines, airborne armament, and so on.

The driving force among the research leaders of this organization was Adolph Baeumker. He was responsible for the lavish way in which the various research institutes were set up. He understood that for successful work on new ideas, one needed well-equipped laboratories, capable personnel, proper living conditions, good co-ordination between scientists, and a minimum of administrative interference. He privately published a book about the problems of government-controlled research, which, except for being long-winded, contains some well thought out plans and observations. It is noteworthy that this book, published in 1943, contains no reference to Nazi party policies or doctrines. Under Baeumker, the organization developed closer liaison between research, development, manufacture, and testing of equipment. Some problems were farmed out to the research laboratories of industry and academic institutions.

It was the ideal of the university research men to have their work sponsored by the Air Force, for then they could be assured of enough material, men, and money. Several did succeed in interesting this powerful group, even though their work was far removed from practical Air Force needs. But a lavish organization of this nature could afford to go in for a few side lines and, besides, they were probably ambitious to co-ordinate all German research under their auspices.

In several phases of aeronautical research the Germans were well ahead of the Allies. We need mention only their jet-propelled fighters, their wind tunnels—especially those for speeds greater than sound—their guided missiles and their V-1 engine. The large number of secret research reports, and secret reports on meetings of the Aeronautical Academy contain a wealth of useful information, well printed, ably organized, and classified for easy consultation.

The German Navy too had a research section which on the whole was properly organized. They, too, kept aloof from other research groups but they farmed out several projects to industrial and university laboratories. Their facilities and the number of their personnel were on a much smaller scale than the Air Force's, but they employed several very able civilian scientists. Excellent work was done towards the end of the war on radical U-boat and torpedo improvements, which might have been fatal to Allied shipping if they had come earlier.

Among the German Navy officers themselves, there were some, however, whose confidence in science and technology seemed not to have been very high. They preferred the more mysterious occult ways of waging war, probably inspired by Hitler's use of astrologers. Instead of concentrating on some kind of radar to locate our convoys, they thought up a much simpler scheme. They used a map of the Atlantic on which they placed a miniature metal ship. A pendulum on a thin thread was suspended over that ship and when the pendulum moved it was supposed to indicate the whereabouts of an Allied convoy. Anonymous letters complained about this kind of warfare. It was alleged that certain officers, who really knew better, misused this superstition of their superiors to get an easy job, spending most of their day having a good time at sea-side resorts. The admiral in charge replied to the Berlin office, that these researches had been undertaken to settle objectively whether the method was useful but that the results were such that the work had recently been discontinued.

The Germans were not the only ones who believe in superstitions. There are business men among us who live according to their horoscopes and colonels who believe in ghosts. Such beliefs make life much simpler and obviate a lot of difficult decisions and responsibilities. It is so much easier to blame the occult pendulum if the convoy is not found than to accuse the admiral in charge of Naval Intelligence. During the war the newspapers in England and in several other countries were reduced to the size of a small pamphlet, but whatever interesting and important sections were omitted, they never cut out their little astrological corner.

The civilian research, which contained all the great brains of German science, should have been the best organized and the most important. But this was not the case.

All academic work and various research institutions, such as the Bureau of Standards, came under the Ministry of Education. The head of this ministry, whose name was Bernhard Rust, was a weak, insignificant Nazi, who did not understand what was wanted of him. He often arrived at important meetings under the influence of alcohol, behaved boorishly and in general made an unfavorable impression. It is said that during the first World War, when he was somewhere near the front, he signed the letters to his children with "Euer Heldenvater"—"your heroic father."

It was this same Rust who had, before the war, made his henchman, the "sturer Nazi" Mentzel of the many titles, executive director of the Reich's Research Council, and thus chief of all research.

Rust was probably the most insignificant man in Hitler's cabinet; he had neither the understanding nor the capacity to promote the interests of German science. But then, from the very outset, the Nazi regime had mistrusted academic science and the scientists. Inevitably this hostile, or at least negative, attitude was detrimental to the advancement of science and the training of promising new research workers. In physics, for example, there was a strong prejudice against Einstein's theory of relativity, and indeed all abstract theory. Mindful of the danger to physics inherent in such a prejudice, the more enlightened physicists had to waste valuable time trying to convert their Nazi-minded, inferior colleagues. In November, 1940, they held a meeting in Munich at which they succeeded in getting an official agreement on the following points:

"1. Theoretical physics is an indispensable part of all physics.

"2. The special theory of relativity belongs to the experimentally verified facts of physics. Its application to cosmic problems, however, is still uncertain.

"3. The theory of relativity has nothing to do with a general relativistic philosophy. No new concepts of time and space have been introduced.

"4. Modern quantum theory is the only method known to describe quantitatively the properties of the atom. As yet, no one has been able to go beyond this mathematical formalism to obtain a deeper understanding of the atomic structure."

This credo was signed by a dozen physicists, half of them reasonable fellows, the other half belonging to the fanatical opposition.

It was not enough. Two years later it was necessary to call another meeting, this time in

Seefeld, in the Tyrolean Alps. The meeting was attended by thirty scientists, some of whom were representatives of the Nazi government's teachers organization. Again they arrived at a compromise; they noted that "the apparent difference of opinion was almost entirely due to misunderstandings."

A lengthy official report was concocted, chiefly by compromisers Sauter and Von Weizsäcker; in general, the report was similar to the credo of 1940. It emphasized the fact that "before Einstein, Aryan scientists like Lorentz, Hasenohrl, Poincaré, etc., had created the foundations of the theory of relativity, and Einstein merely followed up the already existing ideas consistently and added the corner stone." The final paragraph, as toned down by the diplomat Weizsäcker, reads:

"At the Seefeld meeting the opinion was expressed, however, that one must reject the forcing (Ausschlachtung) of the physical relativity theory into a world philosophy of relativism, as has been attempted by the Jewish propaganda press of the previous era (Systemzeit)."

The Alsos Mission found a partial stenographic account of some of the discussions which took place at the Seefeld conference; it was a pretty sickening performance. One wonders how a great man like Heisenberg could stomach the hopelessly stupid opposition, an opposition all the more embarrassing in that it came from so-called fellow scientists, and not from Nazi politicians or other outsiders. Heisenberg, as we have already seen, had less trouble convincing Himmler himself of the necessity of teaching the theory of relativity.

The genuine physicists had a time of it not only in keeping ideas out of the fumbling grasp of the charlatans, but research funds and apparatus as well. They fought successfully a fantastic scheme promoted by the X-ray scientist, Schieboldt. This gentleman wanted to use a new high-voltage X-ray machine, the so-called Betatron—an American invention—against Allied bombers. His idea was to use X-rays to burn the crews out of the bombers. For this pipe dream, he had actually obtained the support of Air Marshal Milch, not to mention his diverting important apparatus from the uranium project.

Another nonsense project aimed to find out whether two infrared rays, intersecting just under the proper angle, could explode the bomb load of Allied planes in the air. This project bore the mythological code name of "Hadubrand"; it was mythological indeed. Naturally, the true physicists knew the idea was senseless, but a few pretended to go along with it since it provided an opportunity for some interesting research on molecular structure.

Aware of the important rôle science could play

in the war, several patriotic research workers offered their services to the government. Those who approached Rust were quite unsuccessful. But even those who got in touch with the Army and Navy directly were not received with open arms. A small group of competent scientists which did some promising work on mines and torpedos for the German Navy at the beginning of the war, soon was disbanded because of lack of work and support.

The fiasco of the U-boat war, the successful defense of England against air-raids, and many other symptoms made it finally clear, even to the stubborn Hitler clique, that the Allies were using scientific equipment to a great advantage. It was not until late in 1942 that they finally decided it was necessary to use Germany's scientific and technological power to a larger extent.

How to mobilize German science, seemed at first a simple problem. Everyone agreed that German Air Force's research under Goering had been immensely successful. It was erroneously believed that this was due to Goering's insight in such matters and to his political power; actually it was due to the ability of its Committee of Research Directors (Forschungs Führung). It was therefore decided to take the Research Council away from Rust and place it directly under Goering.

The German scientists were very hopeful when this change took place. Now at last they would be able to help the war effort; they would get money, men, priorities, materials and, what was more important to them, a real say in matters of scientific warfare. They did get more money, Goering gave them 50 million marks, but they didn't know how to spend it and returned half of it at the end of the year. Except for minor changes in personnel and more impressive letter-heads on their stationery, the change-over from Rust to Goering was no improvement. The basic reason for this failure was that Goering took over Rudolph Mentzel, who continued business as usual. He remained the acting-president of the Research Council and presided at their very rare meetings. Goering never appeared and seldom took an active interest in this unwieldy, badly managed organization. He personally did sign some of the more important directives but most of the policies were set by the incompetent Mentzel while the papers were signed in colored pencil with Goering's name by a stooge on the Reichs-marshal's staff.

Meanwhile, the U-boat warfare deteriorated. On May 25, 1943, the German Navy arranged a meeting of about 200 top scientists. They listened to a couple of very elementary lectures on U-boat tactics. This was necessary because most of the scientists hardly knew how a U-boat functioned. Next they heard a simple lecture on radar and then everyone went home; no plans for action had been made, no suggestions for work. This was the first and last time the civilian scientists got information about operational problems of the Armed Forces. The result was of course nil. When an optimistic scientist, a famous aeronautical expert, inspired by this meeting, sent the Navy some concrete suggestions and plans, he was rebuffed by them. They replied that it was none of his business and that the matter was well under control.

Some of the scientists gave up hope of ever playing a rôle in the war. They were, however, well aware of the importance of science in the potential strength of a country, economic or military. Thus their next concern was to see that Germany kept its supposed dominating position in science. They were afraid that "after the victory" Germany might find herself to have nevertheless lost the battle of the laboratories. To be sure, they were still ahead of the Allies, they said, although some doubt must have begun to enter their minds at that time. But now they used the slogan "The War in the Service of Science," instead of the reverse.

The president of the German Physics Society, Carl Ramsauer, had some very clear ideas about what was going wrong in Germany. He had earlier done outstanding research work on atomic structure and was the director of research for the great electrical concern, the A.E.G. He understood that physics was a basic science for all other natural science as well as for technological applications. He realized that the gap between application and pure science in this field had been narrowed almost to the point of non-existence. He was also aware that the most recent advances in physics, those on uranium, were sure to have tremendous repercussions in economics and in the art of warfare.

In spite of his influential position, Ramsauer was unable to get the authorities to listen to him, with the exception, significantly enough, of the Air Force people. He lectured before the German Academy for Aeronautical Research in April, 1943, and was strongly supported by Baeumker. Ramsauer pointed out that "physics is a weapon of possibly decisive importance in the military and economic struggle of nations." He maintained that the United States and Great Britain had taken the lead away from Germany in this science and that, surprisingly, the United States even excelled in the organization of science, an art that was supposed to be a German monopoly. He discussed, as sharply as he dared, the reasons for the decline of German physics, "a first-rate military power factor." Concluded Ramsauer:

> "Three thousand physicists may perhaps decide the war; 3000 soldiers less does not weaken the Armed Forces. . . . We do not need to fear Anglo-Saxon physics if we are able to utilize the full potentialities of our universities and engineering schools. But if

we are unable to do this, then God have mercy on us."

Ramsauer's talk was printed in a secret publication, but the last sentence was censored and did not appear in print. His aeronautical colleagues seem to have agreed with him. But they pointed out that the necessary reorganization would take years; in the current crisis, they would have to get along by improvising.

XII

The Uranium Club

OTTO HAHN, the German scientist, discovered the principle of uranium fission in December, 1938. Within a few months the entire world of physics had realized the importance of Hahn's discovery; physicists everywhere sensed its significant implications. They called the attention of their respective governments to the importance of this discovery, and informed their governments that it might lead to super-explosives or to unheard-of production of power. In Germany, some scientists communicated with the high command; others wrote to the Ministry of Education, which was at that time in charge of all university research.

As scientific adviser to Army Ordnance, Professor Schumann made immediate preparations for secret research into the uranium problem with a view to producing the super-explosive. But he himself was only a second-rate physicist, and his helpers were not much better.

In the Ministry of Education the man in charge of physics was the president of the Ministry's Bureau of Standards. He bore the surprisingly non-Aryan sounding name of Abraham Esau, and while he had some knowledge of electronics, he could hardly be called a physicist. He had attained his post mainly by being an ardent Nazi.

Esau called a meeting of half a dozen physicists in April, 1939—Professors Joos, Hanle, Geiger, Mattauch, Bothe, and Hoffmann. That was the beginning of the "Uran Verein"—the Uranium Club. The professors were pledged to secrecy and promised the full support of the Ministry. The uranium problem was to be made the co-operative project of the most important physicists in Germany and headquarters were to be set up in the Kaiser Wilhelm Institute for Physics in Berlin. All available uranium was to be earmarked for the researches to be conducted in the various laboratories.

Thus the newly formed Uranium Club proceeded with arrangements, quite unaware that Schumann had a group working on the same problem for Army Ordnance.

In the summer of 1939 two members of the Uranium Club came to the United States, where they had been invited to lecture. They never mentioned anything about what they were doing in Germany, but apparently kept their eyes and ears open for any developments over here. However, at that time very little had been accomplished. The only place where some work was be-

Der Präsident
der
P. T. R.

Den 13. 11. 1939.

Lawrence et al.

Verfasst am 10/9.74

 Im Januar d.J. erschien die erste Veröffentlichung über den Effekt von Hahn KWJ-Berlin, der zahlreiche französische, amerikanische und englische Veröffentlichungen folgten. Im April gab ein Brief des Professor Joos dem Reichserziehungsministerium Anlass, eine Sitzung einzuberufen, deren Vorsitz mir als Fachspartenleiter im Reichsforschungsrat übertragen wurde. An dieser Sitzung nahmen die Professoren J o o s , H a n l e , G e i -g e r , M a t t a u c h , B o t h e , H o f f m a n n teil ausserdem der zuständige Referent des Reichserziehungsministerium Dr. D a m e s. Anschliessend an das Referat von Prof. Joos und Hanle über den bisherigen Stand im In-und Ausland wurde die Frage diskutiert, inwieweit sich ein derartiger Versuch durchführen liesse. Es wurde beschlossen, die Sache als Gemeinschaftsarbeit der bedeutendsten auf diesem Gebiet arbeitenden Physiker in Angriff zu nehmen. Die Federführung wurde mir übertragen. Es wurden zunächst Untersuchungen eingeleitet um festzustellen, wie rein das Versuchsmaterial (Uranoxyd) sein müsste. In Göttingen wurde deshalb Material von Kahlbaum untersucht und gleichzeitig mit dem Reichswirtschaftsministerium, das für die Grube in Joachimstal zuständig ist, über die Bereitstellung von Radium verhandelt. Die Beschaffung des Urans in der notwendigen grossen Menge war schwieriger. Es gelang unterdessen durch tatkräftige Mithilfe des Wirtschaftsministeriums diese Menge sicherzustellen.

 Von diesem Material wurde sofort eine Probe zur Feinanalyse nach Göttingen an die Stelle gesandt, die bereits die früheren Analysen gemacht hatte und über grosse Erfahrungen verfügte. Da der betreffende Herr plötzlich eingezogen wurde, habe ich die Durchführung der Analyse bei der Reichsanstalt sofort in Angriff nehmen und durchführen lassen.

 Soweit war neben allgemeinen Vorbereitungen die Angelegenheit am 4.9.39 gediehen. Ich hielt nun den Zeitpunkt für gekommen, die Angelegenheit Ihnen, Herr General, vorzutragen und um Unterstützung zu bitten, die mir auch zugesagt wurde. Herrn Ministerial-
b.w.

Memorandum mentioning the foundation of the Uranium Club

ing done was at Columbia University, where Enrico Fermi was conducting researches that were to lead, eventually, to his successful production of a chain reaction in a uranium pile. But that summer, Fermi was teaching at the University of Michigan, and it was at my house that Werner Heisenberg, leading German physicist, met him. This was probably the only occasion the Germans had to get any authentic information of what we were doing. The next year the *New York Times* published a rather inflated article about some of the research being done at Columbia under the auspices of the Navy. We found a copy of that *Times* article later in the German files. But that was all.

One of the two scientific visitors, we learned later, had been ordered to report regularly to the various German consulates as he travelled throughout the country as far west as California. Documents uncovered by the Alsos Mission in Germany showed that the Nazis had indeed been keeping an eye on him. There was a good-looking woman on the same ship with him coming over and, interestingly enough, she was right there near him on the trip across the country. That she had something more than romance in mind would seem to be indicated by the fact that she turned up later as an employee of the notorious German consulate in San Francisco. And that she was a personage of some importance would seem to be clear from the fact that later in the war, she was sent to join the German consulate in the spy-infested city of Lisbon.

However that may be, the two visiting scientists returned to Germany with the conviction that we were sound asleep in the United States, that we were doing practically nothing on the uranium problem—just ordinary routine, purely academic nuclear physics. It was a good state of mind for our enemies to have. It lasted them and comforted them until the end of the war.

Shortly after the two scientists returned to their native country, war broke out. Then the Uranium Club discovered that they could not get the uranium which had been promised them because the group in Army Ordnance had earmarked it all for themselves. What made this even more irritating was that the academic scientists considered Schumann and his group far below their level. They thought it outrageous that such men should be given so much power, and felt certain that they would never succeed in their researches. The quarrel was finally settled by an agreement to share the material and to try to co-operate by an exchange of information between the two groups.

But there was still a third group interested in the uranium problem. In Berlin, a very able technician, Baron Manfred von Ardenne, had a private laboratory where he designed and built instruments for industrial and academic laboratories, such as electron microscopes, cyclotrons, million-volt accelerators, and the like. Von Ardenne was not a physicist in the German academic sense, but he was a first-rate experimenter; a designer and builder of important laboratory apparatus, and a successful business man. He found out that the Postal Department had a research section with a large budget that was not being used. Contacting Ohnesorge, the gullible Postal Minister, he told him all about the wonders of atomic power and explosives.

And so it came about that Von Ardenne's Berlin laboratory was made a branch of Postal Research, and Ohnesorge, at a cabinet meeting, informed Hitler about the uranium bomb.

"Look here, gentlemen," said Hitler, "while you experts are worrying about how to win this war, here is our Postal Minister who brings us the solution."

Ohnesorge is supposed to have taken this remark seriously, although it is thought that Hitler was pulling his leg. One thing is certain, however. For a time the technician Baron Manfred von Ardenne was the official expert on nuclear physics to the Nazi government. Even today the academic physicists refer to this as one of the severest insults they ever received from the government, and the reason for some of them becoming anti-Nazi. "If only the government had taken the true scientists into its confidence instead of those charlatans like Von Ardenne and Schumann," they complained to us on the Alsos Mission.

When we finally caught up with some of these scientists, their first reaction was one of surprise and jealousy that it was possible for American academic scientists to work so closely with the United States Army. As one German colleague, whom I had known in the pre-war years, said to me when I entered his captured laboratory: "Well, Goudsmit, what a surprise to see you here, right in front with your Army. If only we Germans had had the confidence of our Army, and so much influence, then . . . you would not be here now."

Cut off from the Army and from industry, and headed by an incompetent Nazi like Esau, the academic scientists made slow progress on the uranium problem. The true physicists could not give Esau their confidence. Unfamiliar as he was with this field of science he could not fail to provoke resentment among them by his insistence on making decisions himself. Indeed, it is said that he interfered with uranium research rather than guided it. Some indication of his sagacity may be surmised from the fact that, in his reports on the progress of the work, he put first results obtained by his Bureau of Standards laboratory, which had nothing whatever to do with uranium research.

The real brains of the project was Werner Heisenberg. Heisenberg is a man of ideals, but ideals distorted by extreme nationalism and a fanatical belief in his own mission for Germany. He had declined many offers of first-class academic positions in the United States, invariably giving as his reason, "Germany needs me." He was convinced that some day the Nazi power would wane and that he would then be needed more than ever to save German science and further its progress.

As stated previously, Heisenberg had succeeded in convincing the Nazis that modern physics was necessary for Germany and essential for the German war effort. This was considered a great triumph. Its net result was probably somewhat less political interference, but this did not mean that the incompetent men who had been appointed professors of physics in the universities, for political reasons, were now dismissed. They remained at their posts and continued to give inferior instruction. Heisenberg's success can, therefore, hardly be considered a triumph for his reputation. That the Nazis were still far from convinced is clear from the fact that as late as November, 1944, the Gestapo wrote to Professor Von Weizsäcker asking him, as a star pupil of Heisenberg, to give his views on German physics and theoretical physics in general in its relation to the German war effort.

Of course, Heisenberg remained the leading spirit in Germany's uranium project. Its policies regarding scientific research were entirely dominated by him; his word was not to be doubted. But the Führer principle does not work very well in scientific projects, which are essentially collective endeavors and depend on the critical give and take of many minds and viewpoints. Had Heisenberg considered himself, had he been considered by his colleagues, as less the leader and more the co-worker, the German uranium project might have fared better.

By the beginning of 1942, German scientists began to see that this project would have to be set up on a larger scale. The first thing to be done, they decided, was to bring the matter to the attention of high government officials. To this end, in February 1942, they arranged a scientific meeting. In our Intelligence reports we later referred to this meeting as the "coming-out party" of the German uranium project.

The meeting was sponsored jointly by the Minister of Education and the Chief of Army Ordnance. A secret letter of invitation was sent to the highest military, naval and government officials—to Himmler and Speer, to General Keitel, chief of staff, and Admiral Raeder, chief of the Navy, to Reichsmarshal Goering, and Nazi Party boss Bormann, among others.

Said the letter of invitation: "A series of important questions in the field of nuclear physics will be discussed that so far have been worked on in secret because of their importance to the defense of the country." The talks to be given and the

Vortragsfolge

der 2. wissenschaftlichen Tagung der Arbeitsgemeinschaft
»Kernphysik« (Reichsforschungsrat — Heereswaffenamt)
im Haus der Deutschen Forschung,
Berlin-Steglitz, Grunewaldstr. 35,
am 26. 2. 1942 um 11 Uhr

1. Kernphysik als Waffe   Prof. Dr. Schumann

2. Die Spaltung des Urankernes   Prof. Dr. O. Hahn

3. Die theoretischen Grundlagen für die
Energiegewinnung aus der Uranspaltung   Prof. Dr. W. Heisenberg

4. Ergebnisse der bisher untersuchten
Anordnungen zur Energiegewinnung   Prof. Dr. W. Bothe

5. Die Notwendigkeit der allgemeinen
Grundlagenforschung   Prof. Dr. H. Geiger

6. Anreicherung der Uranisotope   Prof. Dr. K. Clusius

7. Die Gewinnung von Schwerem Wasser   Prof. Dr. P. Harteck

8. Über die Erweiterung der Arbeitsgemeinschaft »Kernphysik« durch Be-
teiligung anderer Reichsressorts und
der Industrie   Prof. Dr. Esau

The lecture program for the "coming-out party" of the German
uranium project.

names of the men who were to give them might well have interested a gathering of scientists, but they were bound to be boring and, indeed, unintelligible to the outsider, no matter how high his rank. The program started with a lecture on "Nuclear Physics as a Weapon" by Professor Schumann and ended with a lecture by Abraham Esau on the "Expansion of Nuclear Physics Projects," calling for greater governmental and industrial participation. These two talks were no doubt routine affairs that army and navy officers could understand. But in between were six highly technical, first-rate lectures by Hahn, Heisenberg, Bothe, Geiger, Clusius and Harteck, who were certainly top experts in the field.

The scientists were not content to promise government officials a boring day of technical talks. They wanted to do the whole thing up brown

Versuchsessen

anlässlich der wissenschaftlichen Tagung
des Reichsforschungsrates am 26.2.1942

Vorgericht	Verschiedene Wurstarten (Keimwurst)	mit Roggen und Soja angereichert
Zwischengericht	Brühe mit Bratlingsklösschen und Salzstangen	Bratlingspulver mit Hefe angereichert mit synth. Fett gebacken
Hauptgericht	Schweinebraten Gemischtes Gemüse Salzkartoffeln	feingefrostet feingefrostet mit synth. Fett gedünstet
Nachtisch	Obst Kaffee Gebäck	feingefrostet mit synth. Fett gebacken

"Experimental repast"

170

and so they offered to feed their guests as well. Since it was a scientific gathering they thought the officials might be further impressed if they were given a scientific meal. Thus the menu was titled "Versuchsessen," the best translation of which is probably "experimental repast." It consisted of various deep-frozen and enriched dishes, baked or fried in synthetic fats. All details were clearly indicated on the menu.

The lectures promised a trying enough day; the meal was enough to scare anybody away— anybody but scientists. No wonder the big shots didn't show up. Instead, they sent polite letters, asking to be excused. "You will understand, of course," writes General Keitel, "that I am too busy at the moment and therefore have to decline. I shall let myself be informed about the results, however, and wish your meeting success." Himmler writes that unfortunately he would be out of town on the day of the meeting. Admiral Raeder regretted that he could not attend in person, but promised to send a representative. The coming-out party, in short, was a complete flop. It contributed nothing to the support of the uranium project.

The war was not ending so quickly as the Nazis had hoped and, accordingly, in June, 1942, the Reich's Research Council was taken out of the hands of the Ministry of Education and placed directly under Goering. Six months later Schumann's Army Ordnance checked out of the pic-

Der Chef
des Oberkommandos der Wehrmacht

Berlin W 35, den 21.2.1942
Tirpitzufer 72-76.
Fernsprecher: 21 81 91.

Sehr verehrter Herr Reichsminister !

      Für die liebenswürdige Einladung zu der wissen-
schaftlichen Tagung im "Haus der Deutschen Forschung" am
26.Februar spreche ich Ihnen meinen besten Dank aus.

      So grosse Bedeutung ich auch diesen wissenschaft-
lichen Fragen beimesse, so werden Sie andererseits Ver-
ständnis dafür haben, daß ich Ihnen infolge meiner dienst-
lichen Belastung leider eine Absage erteilen muß. Ich werde
mich über das Ergebnis unterrichten lassen und wünsche
Ihrer Arbeitstagung vollen Erfolg.

             Heil Hitler !

An den
Präsidenten des Reichsforschungsrats
Herrn Reichsminister R u s t
<u>Berlin = Steglitz</u>
Grunewaldstraße 35.

Chief-of-Staff Keitel's refusal to attend the "coming-out party"
of the German uranium project.

ture and their chief experimenter, Diebner, together with all equipment and materials, was turned over to Goering's outfit.

On the whole, the change was beneficial to the German uranium work. It meant unification, even though rival groups now had to work together. But especially the fact that Goering was now the boss of the show—if in title only— boosted the morale of the scientists. The brilliant success of the Air Force's research under Goering was taken to be a good omen. But Mentzel was still there in administrative charge of the Reich's Research Council and Esau remained in charge of nuclear and general physics—and they had grown no less incompetent because they were now working under Goering. Esau simply was not acceptable to the nuclear physicists.

Meanwhile research on the uranium pile was being conducted in several places. The principal work was done at an excellently equipped, underground, bomb-proof laboratory at the Kaiser Wilhelm Institute for Physics in Berlin. (*See* frontispiece.) Experiments were also being carried on at the University of Leipzig and, under Diebner, at an Army proving ground near Berlin. Important contributions were also made at the Kaiser Wilhelm Institute at Heidelberg.

But the time came when it was impossible to work any longer in the cities because of the frequent Allied bombings. Hahn's Institute for Physical Chemistry in Berlin had already been bombed. The Institute for Physics was still intact, but it was decided to evacuate the principal laboratory to somewhere in the country where the chance of bombings was remote. Diebner moved to a schoolhouse in a little town in Thuringia. Heisenberg and his group moved to the village of Hechingen near the Hohenzollern castle where they occupied a wing of a textile factory, while their pile laboratory was set up in a cave in the near-by town of Haigerloch. Many other laboratories were moved from the cities to this region of southern Germany. Hahn's Institute was in the near-by town of Tailfingen. The primitive setup of these new quarters didn't make the work any easier; nevertheless they continued to work on with a fair degree of optimism.

In Hamburg, Harteck had supervised most of the work on heavy water and the separation of Uranium 235. There, some small-scale centrifuge isotope separations were attempted. Now these operations also were moved, and ended up in a wing of a parachute-silk factory in the little town of Celle, near Hannover.

It was not until the beginning of 1944 that the scientists, with the secret help of Speer, Minister of War Production, succeeded in squeezing out Esau, who became the boss of radar and radio research. The Speer ministry had always shown an interest in the uranium problem. It had the last word in assigning priorities, without which the procurement of materials was impossible.

Entrance to the cave-laboratory, hollowed out of rock, where construction of new German pile was begun. Location is Haigerloch, a picturesque village high above the Eyach river in Hohenzollern region of Germany. Cave housed the atomic model reactor made of uranium cubes. (*Courtesy of Mrs. S. Goudsmit*)

Perrin (*left*) and Lansdale (*leaning on spade*) looking for uranium cubes near Haigerloch. (*Courtesy of Brookhaven National Laboratory*)

The German atomic model reactor made of uranium cubes found
in cave of Schloss-kirche in Haigerloch (1945). (From *Das Kaiser-
Wilhelm-Institut für Physik, Geschichte eines Instituts* by Werner
Heisenberg in *Jahrbuch Der Max-Planck-Gesellschaft zur
Förderung Der Wissenschaften E.V.*, 1971)

Alsos team dismantling "uranium machine" in cave at Haigerloch
(1945). Uranium cubes are in center, surrounded by graphite.
Pitching in are Lansdale and Welch (*top left center*), Cecil and
Perrin (*bottom center*) and Rothwell (*right, kneeling*). (*Courtesy
of Brookhaven National Laboratory*)

Castle-church (Schloss-kirche) in Haigerloch. The uranium laboratory was housed in a carved out, cave-like area at its base. (From *Das Kaiser-Wilhelm-Institut für Physik, Geschichte eines Instituts* by Werner Heisenberg in *Jahrbuch Der Max-Planck-Gesellschaft zur Förderung der Wissenschaften E.V.*, 1971)

Speer's support gave some important phases of the project the highest priority—such as the manufacture of uranium metal by the Auer Company. Esau was replaced by a real, first-class physicist, Walther Gerlach, of the University of Munich. An able experimenter, Gerlach was also experienced in dealing with government and Army officials, as well as with prima donna scientists. He was acceptable to all; even his Gestapo record was favorable, in spite of some earlier clashes. He did much to bring unity among the various groups whose members had thus far been rivals.

Gerlach was at first reluctant to take on his new job, but no doubt felt it was his patriotic duty to do so. To a friend, who pointed out that the war was already lost for Germany, he said: "We must keep what we have got. We must let our physicists continue their work in their laboratories and at the universities. We must give them the best equipment and instruments and above all save as much as we can of men and material for after the defeat. That will be my task and my duty, nothing else." With this in mind, he prevented physicists from being drafted into active service, even into the "Volkssturm," which was supposed to make a last-ditch stand.

In spite of administrative changes and the increased importance given to the uranium project, it remained comparatively insignificant. The total number of scientists working on uranium and closely allied problems was less than one hundred. They lacked the equipment necessary for important preliminary and basic laboratory measurements. For instance, they complained in their reports that there were no cyclotrons in Germany, whereas the United States had twenty of these important machines. They had had to go to Paris to work with Joliot's cyclotron. Although half a dozen machines were planned or under construction in Germany, only the one in Heidelberg, in the physics section of the Kaiser Wilhelm Institute for Medical Research, was completed and used before the end of the war.

Such was the background of German research. How far did they get toward a solution of the uranium problem?

They knew, of course, of the possibility of a U-235 bomb, but they considered it practically impossible to separate pure U-235. One can hardly blame them for this. Perhaps only in America could one have visualized and realized an Oak Ridge, where pure U-235 was produced by the huge combined efforts of science, engineering, industry, and the Army. No such vision was apparent among the German scientists and certainly no such gigantic combination of all forces working on all cylinders.

Furthermore, the Germans never thought of using plutonium in the bomb, which enormously simplified the problem. The existence and probable properties of plutonium, though still un-named, had been mentioned in scientific litera-ture before the war, and in a few German secret reports, but they overlooked the practical phase of this side of the problem completely.

In fact, the whole German idea of the bomb was quite different from ours and more primitive in conception. They thought that it might event-ually be possible to construct a pile in which the chain reaction went so fast that it would produce an explosion. Their bomb, that is, was merely an explosive pile and would have proved a fizz com-pared to the real bomb.

It was this misconception which made the Ger-mans believe that an energy-producing pile was the first problem to tackle. In our case it was the other way around. We discovered that it was easier to make an atomic bomb than an atomic power plant.

Our lingering belief in the supremacy of Ger-man science makes it hard for us to accept the fact that the German physicists could have failed so utterly. There are even scientists among us who still refuse to believe that their German con-temporaries could have made such blunders. For these, it is necessary to quote a few German statements which prove the facts beyond all pos-sible doubt.

When the greatest of modern atomic physi-cists, Niels Bohr, fled Denmark in the fall of 1943, he reported that the Germans were merely thinking of an explosive pile. At that time we thought this meant simply that they had succeeded in keeping their real aims secret, even from a scientist as wise as Bohr.

But a secret Gestapo summary, dated May of the same year, states:

> "There are two technical applications of uranium fission.
>
> "1. The *Uranium Engine* can be used as a motor if one succeeds in controlling the fission of atomic nuclei within certain limits. . . .
>
> "2. The *Uranium Bomb* can be realized if one succeeds in bombarding uranium nuclei suddenly with neutrons. The neutrons released in the fission should not be allowed to escape, but their too-large initial speed must be slowed down sufficiently so that they will again produce further fissions. The process propagates itself like an avalanche.
>
> "Mathematical computations based on foreign data have shown that processes 1 and 2 are technically certainly possible."

In a recent press dispatch Otto Hahn is quoted as stating: "We knew that [plutonium] must exist, but we did not succeed in producing this substance."

Heisenberg, according to the Associated Press, said that he advised the German authorities that "atomic explosives could be produced either by the separation of uranium isotopes or by building a uranium pile." This is a typically careful statement, which makes our scientists believe that he meant "of course" to use the pile to produce plutonium. He never thought of it; the pile itself was supposed to be the bomb.

In November, 1944, an engineer of the SS, Himmler's Elite Guard, who had some contact with physicists in Vienna, complained to his superiors that Germany did not work hard enough on an atom bomb. In a short letter he tried to describe the advantages of such a missile. Gerlach was asked to formulate an answer for the SS officials and he states, in a top secret letter, that the ideas of the SS engineer were in many points closely related to the German uranium project, but, as he wrote:

"Unfortunately the *technical* ideas are not correct. According to all available experimental and theoretical investigations, which agree completely on this precise point, it is not possible to obtain the violent increase in nuclear fission with small amounts of material. I can assure you that we have approached just this problem again and again from different viewpoints. Not even fundamental laboratory measurements on this effect can be performed with small quantities; one needs, on the contrary, amounts of at least two tons or more, which is one of the

Der Reichsmarschall
Des Großdeutschen Reiches

Präsident
Des Reichsforschungsrates

Der Bevollmächtigte
für Kernphysikalische Forschung

W.-Nr.

München 22, den 18.11.1944
Physikal.Inst.d.Universität
Ludwigstr. 17, Tel.: 23 455
z.Zt. Berlin-Dahlem,
Boltzmannstr. 20

Geheime Reichssache !

He rrn
Ministerialdirektor
Professor Dr. M e n t z e l
Leiter des Geschäftsführenden Beirates
Berlin-Steglitz
Grunewaldstr. 35

Sehr geehrter Herr Mentzel !   7197/44

Zu den Erfindungsvorschlägen von Herrn M a t z k a erlaube
ich.mir folgendes zu antworten:

Die entwickelten Ideen berühren in vielen Punkten sehr eng
unser U-Vorhaben. Es sind richtige und unrichtige Vorstellungen
darin enthalten.

Leider ist der technische Grundgedanke nicht richtig. Nach
allen bisher vorliegenden Untersuchungen experimenteller und
theoretischer Art, die gerade in diesem Punkte in völliger Über-
einstimmung sind, ist es nicht möglich, die stürmische Vermehrung
der Kernspaltung mit kleinen Substanzmengen zu erhalten. Ich kann
Ihnen versichern, daß wir gerade dieses Problem aus mehr als einem
Grunde immer wieder angegangen sind. Nicht einmal grundsätzliche
Laboratoriumsmessungen über den Effekt sind mit kleinen Mengen
durchführbar; vielmehr benutzt man mindestens Substanzmengen von
2 und mehr Tonnen, einer der Gründe für die Erschwerung der Bear-
beitung des U-Problems. Hiermit fallen auch die Vorstellungen
des Herrn Matzka über Erreichung größerer Kerntreffsicherheit.

Ich darf Sie darauf hinweisen, daß in Kürze ein Bericht über
die letzten U-Brenner-Versuche vorliegen wird, in welchen sich
wieder die Erforderlichkeit enorm großer Materialmengen ergibt.

Ich bedauere, über die in manchen Punkten gut überlegte Notiz
kein besseres Urteil geben zu können.

Mit bestem Gruß und Heil H itler !

Walther Gerlach's letter showing that the German physicists
did not know how to make a small uranium bomb.

180

causes why the uranium problem is so difficult. . . ."

And thus the German physicists worked earn-
estly on the construction of a uranium pile. If
that succeeded, they thought, the bomb would be
only one step further. At any rate a pile might
produce an energy source which would enormously increase Germany's economic power and
war potential, and this work would at least secure for them their world supremacy in science.

Success did not come easily. The progress was
slow. The rival groups in the early days of the
Uranium Club, the many moves to avoid bombing, the primitive facilities in the evacuated laboratories, the shortage of "heavy water" after the
destruction of the Norwegian plant, all hampered
research. But the principal handicap was probably the lack of vision of Heisenberg, whose ideas
seem to have dominated most of the experiments.

For instance, there was one German physicist,
Fritz Houtermans, who came closest to the idea
of plutonium. In a secret report, written as far
back as 1941, he pointed out that a pile might
produce new materials, heavier than uranium,
which would probably have the same explosive
properties. Moreover, he stated that such new
elements could probably be separated by reason-
ably simple chemical methods. But although his
report was reprinted twice, no one seemed to take
any notice of it. Houtermans was not in the good

graces of the Heisenberg clique; since he did not belong to the inner circle, he did not have to be taken seriously.

Heisenberg and Diebner at first worked independently. Diebner's work was, of course, considered inferior by Heisenberg and his elite entourage. But it so happened that Diebner had thought of a much better construction of a uranium pile than Heisenberg. This must have been quite a shock to the Heisenberg group and forced them into closer co-operation with their rivals.

The German physicists never got so far as to have a working pile, but thanks to Diebner's success they came nearer to it. Early in 1945, very near the end of the war in Europe, they believed that their preliminary measurements had finally proved conclusively that a chain-reacting pile was possible.

In February, 1945, panic broke loose in Berlin because of the approach of the Red Army. Gerlach, pale, excited and depressed, had come to Berlin to remove the stock of "heavy water" and other materials for the uranium project and to take it to Heisenberg in Southern Germany. Gerlach was very much afraid that the precious "heavy water" might be destroyed. He hated to let it escape from his supervision and gave Heisenberg strict instruction to safeguard it and the uranium metal. Part of the uranium he stored in his Thüringen laboratory. Gerlach was excited, because of the recent progress that had been

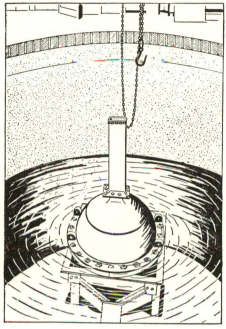

GERMANY'S "ATOM BOMB"

(Above) Drawn from a photograph of the German experimental
"uranium pile," which they believed would make a bomb.

(Right) Diagram for the experimental "bomb" which consisted of
layers of uranium and paraffin.

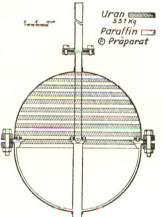

Uran 551 Kg
Paraffin
® Präparat

made. To a visitor who asked what Heisenberg was going to do with all this material he answered, "Perhaps business. Do you know that the U-engine works?" The visitor, who was quite upset by this statement, became inquisitive. Gerlach told him that he had just had word from Heisenberg that the last measurements were in full agreement with the theory. Here the visitor interrupted and said, "But there is quite a difference between a machine ready for work and just the knowledge how it would work *if* it were ready. It usually takes about ten years to put the conception of a scientific idea into technical use," and he mentioned several examples.

"Yes, I know that," answered Gerlach, "but in this case it may be possible to produce the reaction in six or seven months."

"But," the visitor interrupted again, "under present circumstances that means at least a year. The Russians will be in Berlin long before that and all of Germany will soon be occupied by Allied troops. Perhaps there will be a last stand in the Berchtesgaden region, but that means merely a few more weeks until the final fumigation of the Nazis in their caverns. You don't really expect that the work will be continued in Hitler's mountain stronghold?"

Gerlach became more and more uneasy. He cursed the Nazis and the whole war. He wished the Allies would stay where they were so that he could finish his uranium pile. "This is a great triumph," he said. "Think of the implications. Gasoline and radium will be obsolete." Getting more and more excited, he added: "And it is not too late. A wise government, conscious of its responsibilities, could perhaps demand better peace conditions, because we, the Germans, know something of extreme importance which the others don't. But," he added sadly, "we have a government which is neither wise nor has ever had any feeling of responsibility."

Rather cruelly, Gerlach's visitor retorted: "Suppose someone really should be so stupid as to use this for negotiations with the other side. What do you think they would do? Either kill all the physicists, so that they couldn't do any more harm, or keep them behind barbed wire until they have told all they know about the uranium engine or the bomb. But even this may not be necessary. Have you considered that maybe American, British, or Russian scientists know as much or perhaps more about it than you do?"

It is significant that this conversation was reported to the Alsos Mission *before* Hiroshima, before the world knew that the Allies had succeeded in making an atomic bomb. Gerlach's visitor was the German scientist P. Rosbaud, scientific editor of the famous publishing house of Springer. Rosbaud, an Austrian citizen, is among the few who kept their integrity throughout the Nazi regime and the war. His personality and deep understanding gave him the friendship and confidence of all true scientists who came in contact with him and they were many. Everyone knew of his outspoken anti-Nazi feelings and that he tried to keep in contact with Allied colleagues via neutral countries. He was living proof that it was possible to continue unmolested without giving in to the Nazi pressure. He never gave the Nazi salute, never displayed a Nazi flag. There were fortunately a few more among the scientists who acted similarly, notably the physicist Von Laue. These and a few other cases belie the contention that it was absolutely necessary to follow the Nazis in order to be able to live.

XIII

The Gestapo in Science

WHEN Goering took over the Reich's Research Council, a curious character named Osenberg was placed in charge of the newly instituted "Planning Office."

This Osenberg was an obscure Professor of Mechanical Engineering at the University of Hannover, but a good Party member. His technical and scientific knowledge were well below par, but he had supervised some work on torpedoes for the German Navy, which was reported as creditable. He was inspired by a mania for organization and a passion for card indexes.

Osenberg started his career as an organizer of war science with the German Navy. He impressed the authorities with the observation that most academic research facilities were not being used and that the Navy might well take over these places before anyone else got the same idea. With this in view he headed the "Osenberg Committee" which surveyed the various universities. But the Navy soon dropped him, when they found out that he wanted to run and reorganize everything.

187

The Reich's Research Council, which employed him on the rebound, seemed quite pleased with him. At least most scientists were willing to put in a good word for him even after V-E Day. The reasons for this are fairly clear. First of all, Ramsauer's talks and other information had given German scientists an idealized picture of our American organization. What they admired most of all was our much publicized "Roster of Scientific Personnel" and here was this Osenberg, who wanted to set up something just like it for Germany, a complete card index of all German scientists and engineers, and a complete card index of all scientific war projects. There was a still more important reason why they liked him. Osenberg was convinced that scientists should be taken out of the Army and put back in the laboratories on war work. What no one else had been able to do, Osenberg did. He got a Hitler decree passed in December, 1943, dubbed "Osenberg Action," for the release of 5000 scientists from the Armed Forces. "He is the man who really saved German science," the professors said later whenever they were questioned about him.

Osenberg did have a lot of drive. He needed it to get the release decree really executed. He was in continuous quarrels with the Army leaders for his salvaged scientists and at the end of the war only about half of the 5000 had been sent back.

With his extensive personnel files, the boss of the "Planning Office" had a lot to say about the assignment of scientific personnel to the various projects. He had the power to transfer technicians and scientists from one place to another and if one wanted to expand a certain project one had to come to him. He even wanted to supervise the actual programs of research, but in that he was less successful.

From what mysterious source did Osenberg derive his great power? It was no mystery. He was a high member of the Gestapo, Himmler's secret police. The "Security Service of the Elite Guard" (SD der SS), commonly called Gestapo, also boasted of a "cultural" department, Section IIIc, headed by a Wilhelm Spengler. Osenberg was Spengler's right-hand man for the sciences. The function of this section was to enforce the Nazi doctrine at educational and cultural institutions. This was done by means of squealers and investigators who reported directly to Osenberg. All scientific conferences and all important meetings concerning war research and co-ordination were attended by Osenberg's spies. They were also present in all laboratories, whether in the person of a professor or a scrub woman. These spies reported about quarrels between the scientists, inefficiency of the research workers, causes for delays, and other supposed reasons for lack of progress of the war work. In addition, Osenberg collected data on their attitude towards the Nazi doctrine.

This character's Gestapo files were probably

the most revealing documents in his possession. From them we learned who among the leading scientists were considered politically reliable and professionally competent. The physicist Walther Gerlach, the chemists Thiessen and Richard Kuhn are highly praised, but the famous medical scientist Sauerbruch is reported no good as a leader and politically unreliable. Schumann, the chief scientist for the Army, is severely criticized. The able young physicist Gentner, who had been sent to Paris to work in Joliot-Curie's laboratory, is accused of having democratic ideals, probably influenced by his Swiss wife. Indeed Gentner's exemplary behavior during the war at the risk of his own life and freedom completely confirms the poor opinion the Gestapo spies had of him.

Osenberg's agents would investigate research institutes and report on the value and progress of the work done. They gave in some cases more pertinent information than could have been obtained by Allied technical teams.

It was Osenberg's outfit, too, which tried to push Mentzel out of his post as head of the Research Council because of his incompetence and finally almost succeeded, near the end of 1944, after two years of intrigue. A secret report to Goering early in 1943, probably by Osenberg himself, states that "Mentzel is unfit for leadership" and that "a state of chaotic confusion exists in German universities without any coherent discipline."

Mentzel's lengthy reply was quite significant. This "loyal Nazi," as he called himself, suddenly discovered some of the flaws of the regime when he himself was attacked. His defense could almost have been written by an anti-Nazi. He referred openly to the "early lack of recognition by the Nazi Party of the universities, when scientists were obviously regarded as liberal, reactionary, Jewish, or Freemason—in any case, anti-Nazi. This belief was partly justified, and led to a purification which lasted until 1937. . . . Nearly 40% of all professors were dismissed which led to a serious lack of personnel. This could only be repaired slowly; only a limited number of Nazi lecturers and assistants were available to fill the vacancies, and they did not always satisfy the scientific requirements." Mentzel denies the "chaotic confusion" but stresses the indiscriminate drafting of students in the sciences.

A later report, dated August, 1944, and written by one of Osenberg's stooges, critically analyzes the projects sponsored by Mentzel's Research Council, and points out that practically none of them are related to the war effort. Of the 800 projects studied, forestry and agriculture accounted for 70%, physics, only 3%. The only essential problems were on guided missiles. The investigator also complained bitterly about the administration and office routine of the Berlin headquarters of the Reich's Research Council.

The files are in disorder, keys are missing, reports look dirty, and indexing is full of fatal mistakes.

In addition to these undercover reports the "cultural" section of the Gestapo also solicited direct information from the scientists. A secret letter to physicist Von Weizsäcker at Strasbourg, in August, 1944, asks for his views on theoretical physics in relation to German physics and the rôle of theoretical physics in the German war effort. About the same time scientists at the University of Bonn were asked their opinion on "the disintegration of research in the sciences as a result of insufficient governmental guidance."

This letter starts out with the statement that "The advantage which German science and technology possessed before World War I has been wiped out by tremendous developments in this field especially in America." The letter further stresses the important rôle of the scientist in modern warfare and criticizes the Reich's Research Council and similar organizations for having failed to use the German scientific potential exhaustively. It promises a new plan, which intends to remove all obstacles which so far had prevented scientists from contributing effectively to the war effort.

In addition to this, Osenberg sent frequent "Denkschrifte," or memoranda, to the chief of the Nazi Party, Martin Bormann. These were queer looking, immaculately typed pamphlets embellished with underscorings in blue and red ink, beautifully executed, meaningless diagrams, numerous appendixes, cross references, altogether long-winded, pompous affairs in which he aired his complaints. As they referred to almost everything there was hardly a file folder in his office which did not contain one of these pamphlets. It is doubtful whether Bormann or anyone else who received copies ever read the stuff, for occasionally Osenberg complains bitterly about not having gotten any reaction out of them.

One of these memos complains to Bormann that no one in the entourage of Hitler had the courage to tell him that one of his favorite "revenge weapons" against London was a total flop and should be discontinued. The weapon referred to bore the code name "high pressure pump" and consisted of a hundred yard long gun-barrel into which the explosive was fed at intervals along the barrel. Although tests had shown that the thing would not work, thousands of workers were still constructing such installations along the French coast in order not to disappoint the Führer.

Finally Osenberg's dream came true. Late in 1944, Goering was talked into adopting Osenberg's plan. Based on a Hitler decree of June, 1944, ordering the concentration of scientific research towards the war effort, Goering created a

DER FÜHRER

0973/56

4.

Geheime Kommandosache

103/22 gKdos

Führerhauptquartier, den 31. 1. 1945

Die Produktion derjenigen Waffen, die im Notprogramm von mir fest-gelegt worden sind, ist derzeit wichtiger als Einziehungen daraus zur Wehrmacht, zum Volkssturm, Volksaufgebot oder zu anderen Zwecken.

Ich ordne daher an, daß alle im Notprogramm beschäftigten Facharbeiter mit Ausnahme der Jahrgänge 1928 und jünger von jeder Einziehung freizustellen sind, sofern sie nicht vollwertig vor allem durch Fachkräfte stillgelegter Betriebe ersetzt werden können.

Diese Anordnung gilt auch für diejenigen Fertigungen, die als Grund-industrie und Zulieferung zur Fertigung des Notprogrammes notwendig sind (Eisenschaffende Industrie, Zulieferungs-Industrie sowie für die Betriebe, die die Ausrüstungen dafür fertigen: Optik, Elektrotechnik usw.).

Die vorgesehenen oder geplanten Einziehungen müssen unabhängig von diesem Schutz von der übrigen Rüstungswirtschaft aufgebracht werden.

Der für das Notprogramm erforderliche Transportraum ist zu stellen und darf nicht für andere Zwecke beschlagnahmt bzw. abgezogen werden.

Kohle und Energie sind nach Möglichkeit im Rahmen der Einzelanweisungen des Reichsministers für Rüstung und Kriegsproduktion dem Notprogramm zuzuführen.

f. d. R. d. A.

(Saur)

Anlage:
Rüstungs - Notprogramm.

Hitler's technical emergency decree. The scribbles on the right read "Adolph Hitler."

super research council, called the "War Research Pool" (Wehrforschungsgemeinschaft) with Osenberg as the leader directly responsible to Goering, but also keeping his advantageously powerful job in the planning and personnel bureau.

Goering's decree was intended merely as a strengthening of the old Reich's Research Council by putting energetic Osenberg at the top. However, Osenberg extended the interpretation; the new organization was to include also all research facilities of the Army, Navy, Air-Force and industry. He distributed a high-sounding, secret circular on the organization. This included a most complex organization chart, which the recipients promptly dubbed the "railroad switchyard" (Rangier Bahnhof). It looks more difficult than a radio circuit diagram.

It is, of course, superfluous to mention that the research establishments of the Armed Forces, including Goering's own Air Force, completely ignored Osenberg's attempt. The electrical industry was the only one willing to co-operate, but then it had been doing so unofficially for quite some time.

It was now November, 1944. Bombing and the advancing Allied troops had increased the chaos inside Germany. Decidedly, this was not a good time for a new organization to get started. No wonder it never left the paper stage.

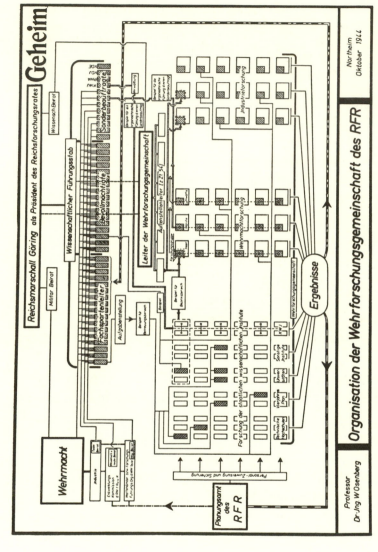

Osenberg's proposed science organization which scientists dubbed "railroad switchyard."

Documents found by the Alsos Mission in Strasbourg in November, 1944, had put us on the trail of Osenberg. We found that his office was evacuated to a little town near Hannover and thought that his files, if found intact, might give us all we ever wanted to know about German war research. We had given the capture of this office a high priority in our plans.

When early in April, 1945, the place was taken, a small group of Alsos military, led by the physicist Major R. A. Fisher and accompanied by physicist Walter Colby of Michigan and chemist C. P. Smyth of Princeton, moved in and captured the whole Osenberg outfit.

As was usual for these Nazis, Osenberg surrendered with all his papers and personnel intact, and offered us his services. The few normal German scientists we encountered always refused to reveal their war work and had hidden or destroyed their secret papers. Not so the Nazis. One reason for their easy surrender was, of course, to save their skins, but this was not the principal motive in a case such as Osenberg's. The truth was that he was so convinced of his own greatness, his indispensability to German science, he was sure the Allies could not govern an occupied Germany without putting him at the head of science. He was greatly impressed by the attention we paid him and even more so when he was taken to Paris.

While the Alsos members were occupied with

a near-by secret nuclear laboratory, some colleagues from Supreme Headquarters moved in and hijacked Osenberg and his menagerie, including all documents. They were put on planes and interned in the previously mentioned "Dustbin" in Versailles. Here Osenberg set up business as usual; he merely had his secretary change the address on his letterhead to "z.Zt.Paris"—"at present in Paris." He was, indeed, very helpful. Various officials asked him for information on technical and scientific programs and he would order his staff to write a very exhaustive report, excellently executed, containing all information on the required subject available in his extensive files and usually ready in an incredibly short time. This strengthened his belief in his indispensability.

A rather stoutish bachelor in his forties, Osenberg was always pleased with himself. People who wanted to get information from him were invariably forced to listen to lengthy talks of his own crackpot ideas on anti-aircraft rockets. It was amusing to observe how he tried to maintain decorum; one of his staff always had to announce his visitors to him. Alsos members felt they could dispense with this rule of etiquette.

Osenberg ruled his staff in a typical German way, by fear. During their internment a revolt broke out. He complained bitterly because his staff had lost respect for Germany's greatness; they would laugh with ridicule when they saw distinguished German internees walk past the office in the château garden on their daily airing. This, he said, was a change in his men which one could not tolerate. He must have sensed that their lack of respect included himself.

One exception was his seemingly sexless secretary, who turned out the best work in the shortest time. She gave the impression of a nervous, mechanical attachment to a typewriter and was under almost hypnotic influence of the "Herr Professor." His male employees, however, who were in many cases more able than Osenberg himself, began to disobey him. They told us how employees who had displeased him had had their draft deferment revoked and had been sent to the front. One sure way to displease Osenberg, it seemed, was to be seen in a movie theater with a girl. They produced the list of former employees and told of the insufficient reasons why each one was fired. Even if the details of their tales were not true, they clearly reflected the abnormal relations between Osenberg and his people.

My friends at Supreme Headquarters, who had hijacked our quarry and who were praised in reports for the discovery of this most significant scientific Intelligence objective, had failed to make a preparatory study of their treasure. They were, therefore, unaware of the fact that some of the most important papers were still missing, namely, Osenberg's Gestapo files and the principal files of the Reich's Research Coun-

cil which had been sent from Berlin to Osenberg's village for safe keeping. I had questioned
Osenberg's men about these papers. They readily
confirmed his relations with the Gestapo, but
claimed that he had burned the papers.

One day, when Osenberg was again pestering
me with his apologies and swearing his loyalty
to the Allies, I became impatient. "I am not interested in your political views," I said, "but only
in the technical information you have. At any
rate, one cannot trust you. You were in charge
of the scientific section of the Gestapo, which
you never revealed to us and you burned all the
relating papers." This unexpected outburst took
him by surprise and he put up a defense by blurt-
ing out, "No, I did not burn those papers, I buried
them and, moreover, I was not the chief of the
scientific section of the Gestapo, I was merely the
second in command!" After that it was a very
simple matter to make him tell where those papers were buried and where the missing Berlin
papers were stored.

Osenberg's signature would be worth a study
by psychiatric graphologists if there are such
experts. Many Nazis seemed to imitate Hitler
and made their official signature into a hiero-
glyph, utterly unintelligible, but easy to fake and
conveying an idea of pathological pomposity.
This habit was especially widespread among
Gestapo officials, although Himmler himself
signed his name very clearly. Compared to their

Teutonic calligraphy, an intricate oriental tughra is a thing of beauty and clarity.

I don't know what became of Osenberg. His Gestapo connections probably put him in the automatic arrest category. At any rate, the revolt of his underlings broke up his dream of future power. He was interned somewhere else and his papers left in "Dustbin" in the care of one of his former slaves.

If Himmler's Gestapo sported a cultural department, his all-embracing SS, or Elite Guard, boasted a whole academy. The SS was a state within a state, with its own government, its own army and, what interests us here, its own science. It was avowedly the last word in Nazi ideals. Its members were supposed to measure up to the ultimates of "pure" Aryanism, fertility, and other ill-digested dogmas, just as its philosophical and religious doctrines were supposed to derive from ancient Teutonic lore. The symbol of the organization— ⚡⚡ —was twice the ancient runic letter S, and not two lightning strokes as has often been incorrectly stated.

During the war the SS had a few technical research laboratories of its own, under the direction of an SS-General Schwab, but these did not amount to anything. They tried some work on heavy water, but soon gave up and sent their "expert" on this subject to the University of Hamburg to continue his work with the legitimate physicists.

The principal "scientific" interest of the SS was ancient Germanic history, with a view to proving the greatness of their Teutonic ancestry. It was for this purpose that Himmler created his own "scientific academy" in 1935, Das Ahnenerbe, or Academy of Ancestral Heritage. Because some of the activities of this strange academy were shrouded in mystery that might just possibly have concealed something really important, we assigned Carl Baumann to make a thorough investigation of the organization for Alsos.

Except for Himmler's letter to hangman Heydrich about the physicist Heisenberg, mentioned in Chapter IX, Baumann did not discover anything connected with atomic research in the Ahnenerbe material. But his report on this academy was most instructive.

In the beginning, the Ahnenerbe was merely a cultural-propaganda section of the SS. But Himmler could never be content with anything so modest. He wanted a full-fledged academy with himself as president. If, as it happened, his academy was duplicating in part the functions of the "culture" ministry of Rosenberg and the propaganda ministry of Goebbels, all the better. This fitted in very well with his method of muscling in wherever possible with a view to eventual control of everything.

Director of Scholarship in the Ahnenerbe was Dr. Walther Wüst, President of the University of Munich, and Professor of Sanskrit and Persian.

His great qualification for his high post in Himmler's academy was that in the early days of Nazism he had defended the "positive" view of Aryan culture in controversies with other professors.

The administrative head was SS Colonel Wolfram Sievers. This psychopathic gentleman was so happy that his name began and ended with an S, he always signed it ⚡iever⚡. He was steeped in Teutonic lore, and while silent about most things he was always willing to talk at length on the subject of runic symbols. Sievers was directly responsible to Himmler and kept him well informed about the activities of his academy. He was also in charge of the organization's publications, books as well as magazines. In addition, he held an important post in the Reich's Research Council. Here he was the understudy of our man Mentzel and had the right to sign all papers. It was another case of penetration on the part of the wily Himmler.

Although, as has been said, the "scientific" work of the Ahnenerbe was mainly historical research to prove that the Nazi ideology was directly descended from ancient Teutonic culture and was therefore superior to all other ideologies, the pseudo-sciences were not neglected. There were divisions for "Genealogy," "Research on the Origin of Proper Names," "Research on Family Symbols (Sippenzeichen) and House Markings," "Spelaeology" and "Folk Lore," not to

203

mention divining rods, and the mysteries of the occult.

Himmler himself was a graduate of an agricultural college and perhaps it was due to this background that he occasionally suggested a sensible research program. Thus he planned an entomological division to study all aspects of insect life and its effect on man. But every so often he could be counted on to come up with something really extraordinary, as may be seen from the following letter he wrote to Sievers from his field headquarters in March, 1944:

> "In future weather researches, which we expect to carry out after the war by systematic organization of an immense number of single observations, I request you to take note of the following:
>
> "The roots, or onions, of the meadow saffron are located at depths that vary from year to year. The deeper they are, the more severe the winter will be; the nearer they are to the surface, the milder the winter.
>
> "This fact was called to my attention by the Führer."

The academy had a few divisions on natural science, although their work was frowned upon by the scientists in the universities. Thus, there was a botany division under Von Luetzelburg, a cousin of Himmler's, who had spent some twenty-seven years in Brazil studying jungle plants and their medicinal properties. There was a section on applied geology which did secret work on the location of oil, minerals and water. Its chief, a Professor Wimmer, spent considerable time with the Army to help them find water in occupied territories. Wimmer is said to have done this by using a divining rod in combination with studies of the slope of the ground.

Among the Ahnenerbe publications was a "Journal for All the Natural Sciences," in which Nazi sympathizers wrote about their "scientific" work. For instance, they had their own pet theory about the structure of the universe, which they called the "Welteislehre," or world ice theory. According to this theory, the inner core of all the planets and all the stars consisted of ice. Not any fancy kind of ice. Just ordinary ice.

In his letter to Heydrich about Heisenberg, Himmler had written: "It would be advisable to bring Professor Heisenberg together with Professor Wüst. . . . Wüst must then try to make contact with Heisenberg, because we might be able to use him in the Ahnenerbe, when eventually it becomes a complete academy, for he is a good scientist and we might make him co-operate with our people of the Welteislehre." It was a suggestion that might well have made Heisenberg shiver.

The Ahnenerbe sponsored archaeological and historical expeditions (Ur-, Vor- und Frühgeschichte!) to foreign countries. These, with true German efficiency, could also serve as bases for military and spy activities. In occupied Russia, the academy's "experts" were on hand to loot the museums of their ancient Gothic art. The only trouble was that the gangs of "culture" minister Rosenberg had been there first. Sievers protested violently against this outrage. He could not see, he wrote, "how these art objects contributed anything to Rosenberg's assignment, which was to collect material for the spiritual fight against Jews and Masons and related world philosophical opponents of National Socialism."

Another example of Sievers' "scientific" interest may be seen in the following letter, written to a Fräulein Erna Piffl, in March, 1943, when the war was at its height.

> "Dear Fräulein Piffl:
>
> "There was a recent report in the press that there is an old woman living in Ribe in Jutland [Denmark], who still possesses knowledge of the knitting methods of the Vikings.
>
> "The Reichsleader [Himmler] desires that we send someone to Jutland immediately to visit this old woman and learn these knitting methods. "Heil Hitler!
>
> " ⚡ iever ⚡"

Unfortunately for the future of science, the records fail to reveal if Miss Piffl's mission was successful.

During the war it was found necessary to add an important new department to the Ahnenerbe —the division for "Applied War Research." This division was responsible for all experimentation on human beings. Since Himmler was in charge of all concentration camps, it followed that any "scientific" work involving their inmates had to be relegated to his academy. When there was a shortage of mathematicians to do computing work in connection with the V-1 and V-2 weapons, a "mathematical section" was made up of concentration camp prisoners who had had mathematical training. They were reported to have done very good work.

But it was rarely that the "Applied War Research" division of the Ahnenerbe operated so humanely. There was, for instance, the notorious "Section H," under Professor August Hirt and Dr. E. Haagen at Strasbourg, who worked on prisoners in the Natzweiler camp. Worst of all was "Section R" at Dachau, where the cruelest experiments were performed by a Dr. Rascher and his very pretty and elegant wife, Nini Rascher, née Diehl. These experiments, which were requested by the Air Force, comprised such investigations as survival after long exposure to extreme cold, and the effects of exposure to extremely low pressures. A thorough study of the complete files of these activities, which were found intact, was made by the well-known Boston physician, Major Leo Alexander of the Medical Corps, and his reports were available to the Alsos Mission.

Aside from the inhuman cruelty of the Rascher operations at Dachau, the work stands out for its absurd attempt at perfection which amounts to a kind of parody of the scientific method. Thus one set of experiments consisted of immersing victims for several hours in ice-cold water until they were almost dead. (Only the hardiest satisfied this particular; most of them succumbed.) Then various ways of revival of the almost dead were tried and their results compared in order to discover the best.

One method was to put the frozen victim in bed with a young woman. With typical Teutonic thoroughness they then tried revival by putting the victim in bed with two women. If time had permitted, they would no doubt have experimented with three, four, and more women, and plotted a learned graph of the results. During the whole experiment the victim's temperature was recorded electrically by means of a thermocouple in the rectum. Major Alexander even found graphs, showing the temperature change until death or revival, marked "rewarming by one woman," "rewarming by two women," and "rewarming by women after coitus."

Although Sievers was in direct charge of all the war research sections of the Ahnenerbe, Himmler himself seems to have taken a personal interest. Most of the letters and reports from Rascher were directed to him. In one of those Rascher requests that he be transferred to Concentration Camp Auschwitz, because it was much colder there and he could cool the victims by leaving them out in the open, naked. Dachau, he wrote, was also too small; his experiments caused some trouble among the other inmates, because "his patients roared while being frozen."

When we interrogated Sievers, he first denied all knowledge of human experiments until we confronted him with some conclusive evidence that he was lying. Even then he seemed only mildly interested and preferred to discuss the prehistoric glory of the Teutonic peoples. He did, however, tell us that his close friends, the Raschers, had ended up as concentration camp prisoners themselves. There were probably several reasons for this, but the one stressed by Sievers was that they had violated the SS code of honor. Nini had had a miscarriage and had substituted another child as her own.

As we have seen, the thoroughness which we admire so much in German science can at times become a parody of science. In the professional library of the Gestapo in Berlin we found a book on "Germanic Symbols." It shows hundreds of runes and other emblems for which it supplies nauseating explanations. For example: "The dumbbell is the symbol for oppositions, counterpart. Birth and death, life and death, old and new year, winter and summer, heaven and earth, the

receiving, the conserving, the generality, etc."
Eminently logical and thorough, this book naturally starts with the beginning of everything, the
sacred dot, "der Punkt." The dot, we are informed,
is "the symbol of all symbols, meaning the beginning and end of all life, the innermost core and
source of power of all formations. It is the symbol
of the germ, but also of the remainder of all
life. . . ."

This type of exaggerated nonsense flourished
especially during the Nazi regime, but it had always been taken more seriously in Germany than
anywhere else because of the pretentious and
pompous way the pseudo sciences were presented.
The very style of German books often made it impossible for a non-expert to judge the soundness of
their contents; compendiums of utter nonsense
were written like learned texts with numerous
footnotes and references, tables and illustrations.
Sometimes good books were spoiled by this over-
thoroughness.

There is an old anecdote that goes the rounds
every five years or so in slightly revised form. It
tells about a group of learned men of different
nationalities who meet in the zoo and are greatly
impressed by the camel. They decide that each
shall write a book about it. The Englishman is first
with a book titled "Camel Hunting in the Colo-
nies." The Frenchman writes about "Le Chameau
et Ses Amours"; the American on "Bigger and
Better Camels." The German, after two years, an-

nounces a "Handbook on Camels"; Volume I—
"The Camel in the Middle Ages," Volume II
—"The Camel in Modern German Civilization."

There is more than just an element of truth in
the last part of this. With some modification, we
found just such a German book. It is not about
camels but about dogs. Not about all dogs, of
course, but only about German dogs, and specifically the German shepherd. The book I am referring to is "The German Shepherd Dog in Words
and Pictures" by Captain Von Stephanitz. It was
first published in 1901 and I have the sixth edition
published in 1921, long before the advent of Hitler. It is definitely one of the best books on the care
and breeding of dogs, but of interest to us here is
the way its informative material is smothered under a weight of pretentious nonsense.

Almost eight hundred pages long, this tome
starts, like so many German books, at the very beginning of things—the creation of the world. To
make it more impressive the creation is introduced
by a quotation of the "Vendidad, the oldest book
of the Zend-Avesta." No doubt every German dog
breeder has this ancient Persian work on his book-
shelf. The first two hundred pages then deal with
the origin of the shepherd dog and its occurrence
in all periods and all over the world—the dog in
China, the dog in ancient Greece, the dog in the
Bible, the dog in Egypt. There is also an interesting section about "the dog and the Jews."

We learn that the ancient Jews despised dogs

and this explains partially "the present contempt for dogs even among Aryan people, which can be blamed on the great influence of Jewish notions, which smuggled themselves in, hidden under the Christian religion." Furthermore, "the attitude of the Jew to the dog is still the same nowadays. . . . Never can the dog have any emotional value to him, never can he devote himself unselfishly. . . . That only a German can do, for 'being a German means to do a thing for its own sake.' [Wagner]"

Among the illustrations in this by no means singular book is a drawing of a dog with the caption "Friendly greeting; after Professor B. Schmid." Of course, a professor was needed to analyze the dog's mood; otherwise the readers could not accept the statement as authoritative.

Finally, German books always have such an excellent index. "The German Shepherd Dog in Words and Pictures" is no exception. I shall list just a few consecutive items from the index of this book to indicate to what extremes German thoroughness can go.

Hund und andere Tiere	Dog and other animals
" und Dienstboten	" and servants
" und Frau	" and mistress
" und Herr	" and master
" und Hund	" and dog
" und Kinder	" and children
" und Spielzeug	" and toys
" und Werkzeug	" and tools

As I have said, Captain Stephanitz' classic on the German shepherd dog was written long before Hitler and Himmler. It helps us to understand that not all the farcical elements in the SS Academy were supplied by Nazism.

XIV

The Efficiency of German Industry

GERMAN industry under Nazi totalitarianism, no more than German science, was the smooth-running, efficient machine some of us had supposed. Here, too, we find that the Nazis had their own super-duper way of getting in each other's road and falling over each other's feet.

Returning to Paris one day after a long field trip, I found on my desk at Alsos headquarters a stack of documents which plainly did not belong there, since they had nothing to do with the work of our Mission. Investigation showed that they had been sent to Alsos through the proper channels from some field headquarters collection center. But no one had requested them or knew why they had been sent to us. It was just another minor Army mystery. When I tried to return the papers to their source, they were politely but firmly refused. The same thing happened later when I tried to turn them over to various divisions of the War Department. Nobody had the least interest in my documents, so there I was, stuck with them.

And yet, although they had nothing to do with

science or uranium research in Germany, they told quite a story—of ruthless bungling.

The documents came from Himmler's "personal staff" and dealt with the vain efforts of the SS gangsters to move German aircraft factories into bombproof caves. The SS moved so swiftly and so efficiently that they were soon involved in serious clashes with both labor and industry, with the result that in six months' time exactly nothing was accomplished. The dossier contains the correspondence on this undertaking with Himmler's office in 1944. The actual writing of the documents was done by SS-Colonel Doktor R. Brandt, of Himmler's staff, but most of them show Himmler's initials written with his familiar greasy green pencil; some have marginal notes by the master as well.

With the ever-mounting crescendo of Allied aerial bombing, the Germans were faced with the urgent need both of stepping up air production and putting industry underground. To accomplish this, Hitler created a "Fighter Plane Committee," called "Jaegerstab," consisting of top men in the Nazi Party, government, Air Force and industry. Accordingly Goering, in February, 1944, sent the following telegram, marked "top secret," to Himmler:

> "I should like to request that you place at my disposal the largest possible number of concentration camp prisoners for work in air-

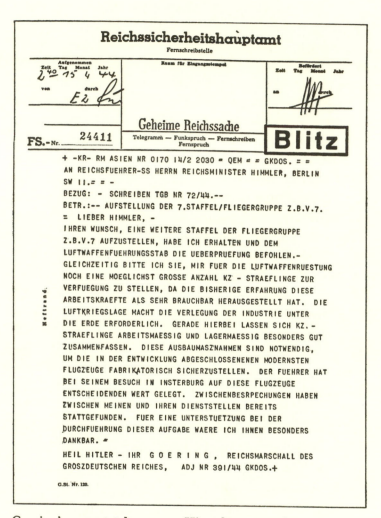

Goering's urgent telegram to Himmler requesting concentration camp prisoners for slave labor.

craft production, since past experience has shown this manpower to be especially well suited for utilization. Present air attacks necessitate the removal of industry underground. Concentration camp prisoners provide an excellent source of labor supply as they are already rounded up and can be quartered in such a way as to make them readily available. As a result of present developments, the removal measure has become imperative in order to insure production of most modern aircraft, models of which have been completed. The Führer, on his last visit to Insterburg, placed decisive emphasis on these planes. Negotiations have already taken place between my department and yours. I should be extremely grateful to you for support in the execution of this project. Heil Hitler! Your Goering, Reichsmarshal of the Great German Reich."

Here was Himmler's golden opportunity to get his fingers into the Air Force's pie. He was not slow to seize it. He replied at once:

"Most Honored Reichsmarshal! Received your telegram of 2/14. Referred your directives to SS-Obergruppenführer Pohl immediately. We will participate with all the resources at our command in the program for removing the factories underground.

217

Heil Hitler! Your, Herr Reichsmarshal, Most Devoted H. Himmler."

SS-General Pohl was the man in actual charge of all concentration camp labor. He submitted a detailed tabulation for eleven concentration camps, giving the number of prisoners from each camp already working on industrial projects, and also the increase he planned. In each case the tabulation indicates the firm and the type of work. All important airplane manufacturers and producers of parts are mentioned—Junkers, Dornier, Heinkel, Siemens and Halske, Messerschmitt, and several more. Each had thousands of prisoners building fuselages, microphones, propellers and other parts, the total being 36,000, which Pohl planned to increase to 90,000. In addition, 100,000 prisoners were required for building the underground factories and putting them in order for production. And yet, in the face of these figures, there were German industrialists who claimed they had known nothing of what went on in the concentration camps!

A master plan was drawn up by Pohl for Himmler's signature. Himmler made only one change. He added a sentence to the fourth paragraph of the report which shows plainly he was thinking of what the deal meant for him in the way of increased power. This sentence I have placed in italics.

The master plan, like all documents submitted

for Hitler's personal scrutiny, was written on a special typewriter in jumbo type. Just why Hitler had to have his documents in special large type is not clear. Perhaps it was his vanity; on the other hand, maybe he was just nearsighted. Once an embarrassing event occurred. Himmler's jumbo typewriter broke down and the boss SS-Colonel Doktor Brandt had to ask his friend Fegelein, who was Himmler's representative in Hitler's immediate entourage, to have a memo retyped before it could be shown to the Führer. This Fegelein, by the way, was married to the sister of Eva Braun, Hitler's mistress. Hitler finally understood that Fegelein had been planted in his influential position to help Himmler to power, and, in the final days of the siege of Berlin, had him executed.

The text of Himmler's basic proposal follows:

"Top Secret
March 9, 1944
"Most Honorable Reichsmarshal!

"Supplementary to my telegram of 2/18/44 I send you herewith a survey of the utilization of concentration camp prisoners in the aircraft industry. From this survey it may be seen that 36,000 prisoners have been put to work up to the present time in the service of the Air Force. It is planned to raise this figure to 90,000.

"Plans for the design and construction of installations are being completed at the present time by the Reich's Air Ministry and the Head of my Office of Economic Administration HQ., SS-Obergruppenführer General of the Waffen-SS Pohl.

"We are co-operating with all the forces at our command.

"The provision of concentration camp prisoners for aircraft work does not exhaust the duties of my Office of Economic Administration, since SS-OGF Pohl and his staff maintain continuous supervision and control of the 'Commands' to see that necessary work tempo is maintained and as a result exercise some influence on output levels. *I may add here for consideration that an increase in our responsibility through a stepping up of the work tempo of the entire industry would certainly lead to increased output.*

"Our own quarry labor forces have been turned over for some time for utilization by the Air Force. Thus in Flossenburg former quarry labor is now working for the fighter plane program in the Messerschmitt factory at Regensburg, on the completion of our stone-cutting shops. After the bombing of Regensburg the opportunity was seized to utilize the prison labor force for the immediate removal of part of the Regensburg plant. Planned expansion there will provide for a total labor force of 4,000 prisoners. At the present time we are producing 900 sets of nacelles and condenser jackets per month and 120,000 parts of various kinds for the fighter plane Me 109 with a labor force of 2,000 prisoners.

"In Oranienburg we have assigned 6,000 prisoners to the Heinkel factory for construction of the He 177. This comprises 60% of the total labor force of the factory.

"The prisoners work excellently. So far they have offered 200 suggestions for plant improvement at Heinkel, which have been used and for which they have received premiums. The prisoner consignment here will be raised to 8,000.

"We have also assigned female prison labor to the aircraft industry. For example, 2,000 female prisoners are working at the present time in the machine shops at Neubrandenburg on the production of bomb releases and rudder mechanisms. The factory here has converted the entire assembly line to prison labor. In the month of January 30,000 releases and 500 rudders and ailerons were produced here. We are increasing the female labor force to 4,000 since their output is excellent.

"In our own factory at Butschowitz near Bruenn we are also producing for the Air Force, although here we use civilian labor.

This factory is turning out 14,000 frames in wood construction for the Messerschmitt-AG, in Augsburg.

"The removal of aircraft factories underground requires an additional consignment of about 100,000 prisoners. Plans for the rounding up of this labor force, in compliance with your memorandum of 2/14/44, are in full swing.

"I will, most honored Reichsmarshal, keep you informed on the further progress of this undertaking.

"Heil Hitler!

"Himmler"

I shall merely give some of the highlights of the subsequent developments, although a detailed study of the whole dossier would be most revealing for specialists on totalitarian methods. There are also a number of interesting marginal issues which we cannot go into here. For example, up to now only the concentration camps were under Himmler, but it was planned to turn some of the military prisoners-of-war camps, the Stalags, over to the SS also. Immediately a quarrel arose between General Pohl and other SS generals as to who should boss these new acquisitions. Himmler finally assured Pohl that he and not the now notorious Kaltenbrunner would get this honor.

The man in charge of the execution of the underground factory plan was SS-General and Lieutenant General of the Waffen-SS Doctor Engineer Hans Kammler. He submitted a beautifully hand-drawn set of colored maps of the proposed installations and an estimate of the progress. In June, 1944, he had three million square feet of underground space ready, which he promised to increase to ten million by the end of the year.

But progress was not easy, as the following report from War Production Minister Speer shows:

"The proposal to erect the second large construction project, not on German soil, because of manpower and material shortages, but on suitable terrain (particularly in respect to transportation and gravel foundation) in the immediate vicinity of the French, Belgian, or Dutch borders met with the Führer's approval, provided the installation is located behind a fortified zone. The greatest advantage in favor of choosing a French site is the availability of the required manpower. In spite of this the Führer urged that efforts be made to find a safer location for this second project, namely in the Protectorate [Czechoslovakia]. Should the required labor force be difficult to supply there, the Führer will personally arrange with the Reichsführer-SS to provide the necessary manpower of about 100,000 men by collecting a contingent of Jews from Hungary for this purpose. The Führer requested that a meeting with him of all participants take place within a short time to discuss details.

"Heil Hitler!

"Speer"

The quarrel between Pohl and other SS generals as to who should control the military prisoners was only one of many revealed in the dossier. Soon there was a fight about space between the aircraft firm Dornier and the airplane engine manufacturer Wankel. The minutes of a meeting held May 11, 1944, contain the following passage:

"In the final discussion, Colonel Geist called special attention to the extremely unfortunate circumstances which offended the Reichsführer-SS [Himmler] and Vice Admiral Heye suggested that in the 5th year of war it was mandatory even for firms with long-standing rivalries to achieve harmonious, comradely co-operation in the interest of the general welfare."

Himmler backed the Wankel firm against recommendation of Marshal General Milch of the Air Force. They had impressed him with a new type of engine which turned out to be a flop. It couldn't even be used for speedboats.

Next in the dossier we find SS-General Engineer Kammler complaining bitterly that the chief of staff, General Keitel, had made no decision yet about the release from the army of skilled workers for mining construction work.

In May there was a clash with the Reich's Labor Front. This organization possessed rest camps, a few of which apparently were ideally suited to house some of the concentration camp workers on the cave project. Engineer Kammler confiscated these camps over the protest of the labor authorities. He wrote wheedling letters to Reich's Labor Leader Hierl and signed them in his most impressive way:

> "Heil Hitler!
> "Your most obedient Dr. Engineer Kammler
> "SS-Gruppenführer and Lieutenant General of the Waffen-SS
> "Commissioner of the Reichsführer-SS to the Reichsminister for Armaments and War Production, Jaegerstab [Fighter Plane Committee]."

Despite this formidable signature, Hierl was not appeased. He sent a furious teletype message to Himmler and War Production Minister Speer:

> "I understand that you are trying to make up for former negligence as fast as possible. I see clearly the importance of your task, but I must demand that you understand one cannot just put the Reich's Labor Divisions on the street. I object sharply to the use of force by your inspection committee. The Reich's Labor Organization, as an armed military state institution, will in future counter any violence with violence. . . ."

Though the papers do not show it, there is little doubt that Hierl lost the battle.

Directly following this trouble with labor, Kammler got into a fight with industry, which led to the arrest of Director Zipprich of the Bavarian Motor Works. This, of course, displeased War Production Minister Speer, who did not think that such procedures by Himmler's gang helped at all to increase the production of airplane engines. But let the papers speak for themselves. The principal memorandum is from a Gestapo security officer named Bischoff to SS-Police Lieutenant General Müller. It tells of a meeting to which Director Zipprich was called to discuss the removal of his plant to an underground tunnel in Alsace. Zipprich, however, did not show up.

"Dr. Zipprich paid no attention to this order, and without offering any explanation of his indisposition sent a deputy, Dr. Kries, head of the BMW-Planning Board. No general preparations necessary for the instruction of the Jaegerstab, such as layouts for setting up the machines, etc., were made, so that the Jaegerstab had to rely on the plans handed over to it just a few days previously by Dr. Kreutz of the Jaegerstab.

"These plans had been sharply criticized by Department Chief Sauer as inadequate in every respect, in the presence of numerous representatives of war plants in Alsace and

Baden, and Dr. Zipprich was openly accused of sabotaging the factory removal program. The entire incident was recorded in the minutes of that meeting.

"As Department Chief Sauer pointed out, Dr. Zipprich's proposal to place the machines at the greatest possible distances from one another, and to provide for enormous rest rooms and locker rooms *inside* the underground site constituted a most irresponsible waste of space, and was nothing short of a crime. The testimony of experts will prove beyond doubt that the BMW firm, under the responsible directorship of Dr. Zipprich, whose deputy constantly referred to him for his orders, unequivocally followed a policy of complying with the removal orders by allocating only as few machines for transferral as they could get away with.

"Having established this, Department Chief Sauer immediately relieved Dr. Kries from his duties as Removal Planner and appointed Dr. Duerr, formerly of the BMW plant, who had been forced out as a result of factory intrigues, as commissar of the entire planning staff of the BMW. Dr. Duerr has already taken up his new duties.

"It must be made clear that by May 1st, 1944, the entire length of 7 kilometers was laid with concrete. Beginning from that date on, at the very latest, mounting of the first machines should have begun. Instead, up to the day of the Jaegerstab's inspection, not a single machine from the Allach factory had been prepared for transportation. Such behavior can only be characterized as the most downright effrontery towards the SS-Führungsstab and the construction firms under its direction, as well as their personnel and workers. With the most willing co-operation of the German workers, working tirelessly day and night, Sundays and holidays, even during the Easter holiday, in recognition of the decisive importance of their task in winning the war, their project was completed, even before the deadline. But Herr Zipprich was entirely unmoved, and with the utmost indifference sabotaged the directives of the Führer in the most unqualified manner. . . .

"After my return on the special train I proposed to Hauptdienstleiter Sauer that this monstrous case not be disposed of in the usual way, simply by removing the responsible individuals from the BMW, but that he give his permission that an example be made right at the outset to serve as a precedent for any future incident that might arise during the countless removal measures that were yet to be undertaken, to insure that those gentlemen of big business who still had a liberal-capitalist orientation would lose their taste, once and for all, for behaving in a similar manner in the future. Department Chief Sauer approved of this proposal, as did the representative of the fighting front who was present at the time, Bearer of the Knight's Cross with Oak Leaf Cluster with Swords and Brilliants, Colonel Gollob.

"My proposal, specifically, is to arrest Dr. Zipprich as quickly as possible, in order to prevent any further attempts at sabotage or influence by Dr. Z.

"I beg to be informed, without delay, on the action taken in this case, that I may so inform SS-Gruppenführer and Lieutenant General of the Waffen-SS Chief Engineer Dr. Kammler.

Bischoff"

Here we see the typical Nazi method of appointing, as the new head of the firm, a certain Dr. Duerr, who had been fired by that same firm. It is at least doubtful if such procedures improve the efficiency of industry. And then Gestapo spy Bischoff shedding tears over the "German workers," who toil tirelessly day and night, even during the Easter holiday, on the construction of the tunnel. He must have forgotten that these "German workers" were concentration camp prisoners.

Another serious quarrel occurred with Army Ordnance. General Pohl complained that they had failed to use certain caves and he denounced the Chief of Ordnance, General Leeb. The latter wrote to Pohl that, after all, the use of caves was none of his business but that Leeb himself was responsible directly to the Minister of War Production. Himmler was furious. He underscored Leeb's "none of your business" passage and wrote in the margin "Authority Mania" (Zuständigkeitsfimmel). Himmler instructed Pohl to keep an eye on this whole matter and to keep him informed.

The few excerpts I have taken from the dossier will suffice to give some idea of how efficiently everything was managed in a country ruled by secret police methods. It is perhaps superfluous to mention that practically no progress was made with the removal of the aircraft plants into caves. Engineer Kammler wrote glowing and optimistic reports which in time became shorter and more factual. According to his last report, dated December 5, 1944, the total number of concentration camp inmates at work in new underground factories was only 9000 in contrast to the promised increase of 55,000. Allied advances had made it necessary to abandon several planned caves too near the Western Front. In short, the total underground area in actual use in December was exactly the same as it had been in June. Wrote Kammler:

> "The deadlines given in former reports are no longer valid under present conditions of difficulties with transport, gasoline, tools, and the lack of police personnel."

Nevertheless, Himmler, in his large, angular German script, marked Kammler's report "Sehr gut." This praise was duly transmitted to Kammler in a note reminiscent of kindergarten days:

> "The Reichsführer-SS thanks you for sending him the list of completion dates of the special plans. He has put the remark 'very good' on your communication in his own handwriting.
>
> "Heil Hitler!
> your
> R. Brandt
> SS-Colonel."

XV

It Can't Happen Here

THE story is told of a Frenchman who was supposed to have said: "My country is always admirably prepared to fight the previous war. In 1914 we fought the war of 1870. And in 1940 we had the Maginot Line which would have served us so well in 1914."

This story is no doubt apocryphal. And yet there is a grain of truth in it which can well serve us as a warning, lest we become too complacent about the outcome of World War II and think we can rest on our laurels. The scientists, perhaps most of all, dread and loathe the prospect of a third world war, but they also realize how foolhardy it would be for us to assume that the present supremacy of American science will continue uncontested, or that, because we have the atom bomb, the future will take care of itself.

If the foregoing chapters have brought to light certain failures on the part of German scientific organization, certain stupidities on the part of German scientists and their government, it would nevertheless be a bit superfluous for us to derive any comfort from such revelations at this late date. It is far more important for us to learn from them how to avoid similar errors on our own part.

World War II proved conclusively that a country strong in science is in a favorable position to defend itself. The lag which once existed between pure science and its applications has all but disappeared. There was a time when engineers were inclined to be skeptical of theoretical science, but today this attitude is far less prevalent. In such fields as, for example, aerodynamics and radio engineering, the results of mathematical investigations can at once be translated into practical new designs.

But pure science has never been a nationalistic monopoly, and one of the most dangerous mistakes the German scientists made was to assume the supremacy of German science. With very few exceptions, such as Professor Ramsauer, president of the German Physics Society, they smugly took it for granted that their work was better than anything the Allied scientists could achieve. They were especially convinced of this with regard to their uranium research. When Rudolph Mentzel reported in 1943 that their own slow progress on the uranium problem gave assurance to the effect that the Allies must be hopelessly bogged down, it was not merely the opinion of one incompetent. Mentzel was voicing the general state of mind that obtained among German scientists. That is why a full year and a half later, in December,

1944, such an able physicist as Walther Gerlach could state: "I am convinced that we at present have still an important lead over America."

Such complacency, fortunately, is not generally to be found among American scientists. Far from being conceited about their success in producing the atom bomb, they are impressed primarily by its ultimate simplicity. They may underestimate some of the difficulties they had to overcome and especially the enormous engineering problems that had to be solved. But they believe any other country that has the raw materials can make a bomb in a few years. The basic principles are well known to any student of modern physics and the engineering difficulties may have various solutions, some perhaps even simpler than ours.

But if there is little complacency among American scientists, the same thing cannot be said with equal confidence of the general public and certain of their representatives in the government. The belief is far too prevalent that we are far ahead of any possible competition because we have the atom bomb. Many people have a childish notion of the atom bomb "secret" which we must be careful not to "give away." In their minds the "secret" consists of a formula or a diagram on a piece of paper, to be swallowed when a potential enemy approaches. Confident that we, at least, are safe from an attack by atom bombs, they would fetter science lest the "secret" leak out and their night's sleep be severely disturbed. If such notions are permitted to become general, our scientific unpreparedness could let us in for a surprise attack compared with which Pearl Harbor would seem a mere broken window.

German science, as we have seen, was severely handicapped by Nazi dogma. By persecuting and exiling all scholars afflicted with the Jewish "taint," Germany lost some of the greatest scientists in the world. In a healthy country, however, such a loss could have been replaced in a relatively short time by outstanding scholars who were followers of the exiled men. This did not happen in Germany because the effect of the Nazi ideology was to make "non-Aryan" sciences like modern physics, unpopular, with a consequent loss of promising students. Finally, the instruction of the few students who dared to study the abstract, or "non-Aryan" sciences, progressively deteriorated. Quite frequently the Nazis appointed teachers who did not even understand what they were teaching. Thus Munich, under the great Sommerfeld, was once the world's most productive university in theoretical physics. When Sommerfeld retired, shortly before the war, he was replaced by a Nazi named Muller, who did not "believe" in modern physics (probably because he could not grasp its intricacies).

Under the Nazis the students suffered gaps in their training which were hard to fill and made them unfit for the serious work ahead. In addi-

235

V e r m e r k

E /11/2/y24

Der Mitarbeiter der "Zeitschrift für die gesamte Naturwis-
senschaft" Dr. Eduard M a y teilte mir am 9.2.1942 in
München mit, das von den Professoren M ü l l e r / S t a r c k
herausgegebene Werk "Deutsche und jüdische Physik" wurde zur
Besprechung auch an "Das Reich" gesandt, "Das Reich" hat die
Besprechung abgelehnt. Man nimmt an, daß die A blehnung, die
übrigens sonst nur noch von der DAZ und der Frankfurter Zei-
tung erfolgte, darauf zurückzuführen ist, daß "Das Reich".
das Buch zur Besprechung an einen Mann der alten Schule ge-
geben hat, der diese Besprechung ablehnte. Das Werk setzt
sich mit der Einstein'schen Relativitätstheorie auseinan-
der und geht wohl auch nicht zart mit den liberalistisch
eingestellten Professoren um. Frage: Können wir bei H a -
g e n uns für die Besprechung im "Reich" einsetzen ? Die
Besprechung würde gern vornehmen: Prof.Dr. T h ü r i n g ,
Wien.

Berlin am 12.2.42
S/Wo

ϟϟ-Obersturmbannführer

Memo by Sievers about a review of a book on "German and
Jewish Physics" written by Professors Müller and Stark and
opposing Einstein's theory of relativity.

236

tion, valuable time was lost on required indoctrination courses. The efforts of the few remaining genuine scientists to remedy this situation were inadequate. Not only did they have to fight the Nazi officials; many of their own colleagues were infected with the same Nazi neurosis. It was indeed a serious indictment of the German system of education that it produced men who could, at one time, have done outstanding work in a narrow field of research, and yet proved themselves to be dangerously unbalanced in their judgment and behavior once outside their specialized rut. Such men can hardly be called scholars, or even educated human beings. They have the characteristics of a machine or a super robot, which performs a certain prescribed task absolutely correctly, but blows the fuse if used for some task other than that for which it was built. Men like Phillip Lenard and Johannes Stark are typical examples of this perhaps peculiarly German species.

Here in America we have no Nazi doctrines to interfere with the progress of science. Nevertheless there are certain symptoms among us that deserve further thought. There are well-meaning but misguided souls who would hamper the progress of medical science by enacting anti-vivisection laws. There is no difference between Tennessee's law against the teaching of evolution and the Nazi ruling against modern physics. We do not select Nazis to teach at our universities, but some racial discrimination does exist. Of course such discrim-

237

ination is not imposed by our government, but faculty members, deans, presidents, and boards of regents can be quite effective in their own way. If such discrimination persists and increases, it may reach the danger point for American science quite as much as Nazism brought about the deterioration of German science.

Again, we have no Nazi dogma operating to make abstract, "non-Aryan" science unpopular with students. Nevertheless, "long-haired" science has always been held in some disrepute by American students, with the result that the number selecting the pure sciences as a career is still far too small. The ideal held up to American youth is a man like Edison, while a great pathfinder like Josiah Willard Gibbs, of Yale, whose theoretical researches in the 1870's founded a new branch of chemical science, is practically unknown. Yet the conquests of present-day chemistry and chemical engineering are inconceivable without the vision of Gibbs.

Since the atom bomb there has been an increase in the number of physics students. But as soon as the glamour of the new wears off, this number will in all likelihood decrease. A higher salary level and increased prestige for teachers, together with a popularization, on the high school level, of the value of pure science would help to remedy matters.

We cannot go on forever living on borrowed scientific capital from Europe. At present, the roster of some of our specialized scientific societies reads like the line-up of a Notre Dame football team. In the future, we may not be able to import an Enrico Fermi, whose work was the key to our atom bomb, or a great aerodynamical theorist like Von Kármán, or the outstanding expert on vibrations, Stephan Timoshenko, and many others. The sources from which they came are dry now and we shall have to produce our own geniuses. There is no time to lose. We must convince our young people that new ideas are more important to their country and the world than new gadgets, even though the latter may bring in more immediate cash.

Despite the German reputation for organization, the failure of their science during the war was due in part to the way they mismanaged the organization of their research. Proof of this can be seen in the fact that the excellently managed Air Force's research produced superior results, while academic research hardly contributed at all to the war effort. Yet the scientists in both groups were equally capable; indeed, many belonged to both organizations.

Scientific work did not advance under the administration of men like SS-Brigadier General Ministerial Direktor Professor Doktor Rudolph Mentzel, and Ministerial Manager Professor Doktor Erich Schumann, assisted, opposed, and spied upon by an Osenberg with his organization charts and card index mania. These men were obviously bad administrators. But even if their paper work had been in order, they still would not have had the confidence of the bona fide scientists.

One of the principal tasks of research administrators is to bring the results obtained by scientists to the attention of the proper government agencies. Conversely, they must make the scientists aware of problems whose solutions would be of value to the Army, Navy, Air Force, or other wartime or peacetime agencies. But a Mentzel was not able to grasp the significance of the research he was bossing, nor was a Schumann able to understand the needs of the Armed Forces. They had not even the vision to delegate these important functions to more competent underlings; with few exceptions, liaison, which played such an important rôle in the Allied organizations, was absent or inefficient in the German setup. Colonel Geist, in Speer's Ministry of War Production, was perhaps the only really capable liaison official who understood the various aspects of Germany's research needs.

It would be wrong to suppose that we in America are immunized against making such mistakes. The importance of science, and the growth of its needs in time and expense, make some sort of unified control necessary. This is obvious in the case of atomic bomb research, but it is also becoming increasingly essential in other fields of research. The vital question is how to choose the men for the controlling posts. In Germany they were selected because they had the confidence of the Nazi bosses. This is not likely to happen here. But it could be just as fatal if an administrative chief were to be chosen for other irrelevant reasons, let us say because he served a number of years in Congress, or because he was a graduate of West Point or Harvard, or because he was a Democrat or Republican or a bank president or a retired ambassador. The only criterion must be the fitness of the man for the job, and fitness means he must have the confidence both of the scientists who work under him and the government agencies with which he must correlate their activities. Of course, compromises are necessary; there is probably no man among us whose appointment would not be criticized severely by some scientists, Army officials, or politicians. But in the final analysis, it should be the scientists who have the last say about the men chosen to administer their projects.

The difficulties which arose when the Atomic Energy Commission was to be confirmed show clearly how near we can come to making the Nazi mistake. The Commission now functioning has the confidence of most of the scientists involved; if they had not been confirmed our present atomic energy project would certainly have deteriorated completely. But American scientists are no yes-men and if, in the future, the Commission's actions should be at variance with the judgment of the scientists, they will not hesitate to make it known.

Leaders who were excellent under war-time conditions are not necessarily the right men for the job in time of peace. General Groves did remarkable work as the head of the Manhattan Project at a critical time when high speed organization of huge novel constructions in difficult places, and the quick solution of other tough problems, were essential for its success. He showed vision and extreme courage to embark upon this gigantic enterprise. But the General is not, as Senator McKellar stated, "the discoverer of the greatest secret that the world has ever known, the greatest discovery, scientific discovery, that has ever been made." It is just as untrue to say that he retarded the work by a whole year, as has been stated falsely in some quarters. But it would seem that the General's great abilities, useful as they were and still will be in many phases of the work, are not of the kind particularly needed now. The destruction of such harmless but valuable research apparatus as the Japanese cyclotrons— which was due to an error by a subordinate officer on the one hand, and, on the other, to the premature revelations of information in the Smyth Report as recommended by the scientists —indicates that a full understanding of all aspects of the problem had not penetrated sufficiently.

The three German errors mentioned so far— complacency, deterioration of interest in pure science, and regimentation in the administrative control of science, are the principal ones that we, too, can make if we are not on guard against them. There were other mistakes the Germans made, but these fortunately are not likely to befall us— unless we import too many German specialists and blindly follow their example. For instance, there was the lack of teamwork between the various groups of German scientists. In an American organization there would have been a place for an excellent technician like Von Ardenne and a hard-working beginner like Diebner, alongside of first-class scientists like those of the Heisenberg clique. In Germany, on the contrary, the various groups had nothing but contempt for each other. Such an attitude, unlike healthy competition, is not conducive to the promotion of successful research.

Another typically German attitude, detrimental to their progress, was the hero worship of individual scientists. The admiration, which his colleagues rightly had for the great physicist Heisenberg, went so far as to prevent them from thinking critically about his work. But no matter how great a man Heisenberg is, the uranium problem is too extensive to be grasped in its totality by any one individual. What is needed for success on such a project is the clash of ideas of a number of outstanding specialists. Heisenberg never hit on the idea of using plutonium, although in principle it was suggested in a secret report by his colleague Houtermans, who did not belong to

243

the inner circle. Yet if that idea had been taken up,
the German uranium project might have gone
much further. As it was, their immediate goal was
only a slow uranium pile and beyond that, their
eventual goal was the erroneous idea of an explosive pile.

Many physicists on our own uranium project
still find it hard to believe that Heisenberg did not
see what is so obvious to them now. But we must
not forget that our scientists, too, started out with
the idea of an explosive pile in mind. There is no
denying that some of the German preliminary
pile work was excellent; but it was too slow and
on too small a scale to be significant.

Finally, there is the ticklish question of secrecy.
Here we can learn very little from the Germans.
By our standards, their security measures on the
uranium project were insufficient. But however
necessary strict secrecy may be in time of war, in
peacetime it has its decided dangers.

If certain basic scientific discoveries are to be
kept hidden from other scientists because of the
need for secrecy, the result can prove more disastrous than the prohibition of teaching Einstein's
work in Germany. One need only consider what
would happen if advanced students in the sciences
were to be taught the wrong things because the
correct findings must be kept secret. Fortunately,
the present policy allows the publication of all basic scientific discoveries and measurements which
were made in connection with the atomic bomb,

but it is exceedingly difficult to know just where to draw the line. The significant thing is that the original secrecy in our atom bomb project was not imposed by any military or government ruling, but arose spontaneously among the scientists themselves. This is the only healthy way to operate. When it comes to secrecy in scientific matters, the scientists are the best judges of what to keep secret—and when.

Too much secrecy stifles the progress of science. It also fetters the training of young scientists. The war seriously interfered with the education of promising students and now, in view of the greatly increased demand, we have a shortage of physicists and other scientists. But how can we produce new scientists if things it is essential for them to know are to be kept secret? How can we attract young minds into fields that are kept hidden from them? Several great discoveries in the physical sciences were made by very young men at the start of their careers. Cut out this primary source of progress, and science may become stagnant within a generation.

Scientists sometimes boast of the absence of international barriers in their work. There are numerous stories of how British scientists worked on German problems, and vice versa, right through the First World War. During World War II astronomical information was regularly exchanged between Allied and Axis astronomers via neutral countries. Secrecy would, of course, eliminate international co-operation in scientific matters. Even worse, it might lead to the division of groups of scientists in the same country, the same laboratory, and the same project, who are not allowed to exchange information.

The inherent desire among scientists to discuss their work is not a mere excuse for bragging. Nor does their unselfish sharing of scientific information come from any superior traits of character peculiar to scientists. The truth is that the pooling of scientific knowledge is essential for the work of each individual scientist and for the progress of science as a whole.

The desire and the need of the scientists for world-wide co-operation should be an encouraging example to a world that is sorely in need of the co-operative spirit. It is only by breaking down the barriers of dogma, mistrust, fear, and secrecy, only by the free exchange of ideas and the widest possible dissemination of truth that science can continue to raise its level, and with it, the level of our civilization.

APPENDIX

An Outline of the Uranium Problem

ORDINARY uranium is primarily a mixture of two kinds of so-called isotopes. The abundant one is called U-238. The one which interests us most is designated as U-235 and occurs only in 0.7% in ordinary uranium.

A neutron is one of the building stones of the nucleus of an atom. In certain modern laboratory experiments neutrons can be freed from various atomic nuclei. If a free neutron happens to collide with the nucleus of U-235, something new happens. The nucleus of U-235 splits into two parts which fly away with terrific speed; in addition a few extra neutrons break off and fly away, too. This process, called "fission," was the basic discovery made by the German physicist, Otto Hahn, in December, 1938.

Thus, if we had a piece of pure U-235 and a neutron collided with one of its atoms, the release of those few extra neutrons would most likely cause other atoms to split up too. This, in turn, would give rise to more neutrons and more neutrons, which again would explode more and more atoms.

It is like the spreading of a fire. The difference is merely one of degree. The amount of energy released in the breaking up of the U-235 nucleus is a hundred million times more than the amount of energy which occurs in the ordinary burning of an atom.

The simplest concept of the bomb, therefore, is to produce a piece of pure U-235. But to separate pure U-235 from ordinary uranium in which it occurs only in 0.7% is an extremely difficult problem. It was done in the United States at Oak Ridge and represents one of the greatest engineering achievements of our time, made possible only by close co-operation among pure scientists, engineers, and the military.

Now we come to the second part of the uranium problem. Suppose we have ordinary mixed uranium. If a neutron collides with U-235, the new neutrons which are produced in the "fission" are not likely to find another U-235. Instead, the much more abundant U-238 will capture those neutrons, and in that case nothing extraordinary happens. However, by a clever arrangement, it is possible to make most of the U-235 fission neutrons find another U-235 instead of being captured by the abundant U-238. This arrangement utilizes the experimental discovery that slow-moving neutrons are much more readily captured by U-235 than by U-238.

The arrangement is as follows: small pieces of ordinary uranium are imbedded in a substance which has the property of slowing down the neutrons without capturing them. The most favorable substance for this purpose is so-called "heavy water"; another possibility is pure carbon. Such an arrangement is called the *uranium pile*.

If the uranium pile is properly designed, this is what will happen: a neutron which collides with one of the rare U-235 atoms will produce a fission of that atomic nucleus. In this fission new neutrons will be produced. These have a fair chance to fly out of the piece of uranium into the heavy water or the carbon and get slowed down there, before flying into a piece of uranium. And because they have been slowed down, they have a much better chance of being captured by the rare U-235 and produce a new fission than to be captured by U-238.

It is all a question of proper balance. If the balance is just right, so that for each U-235 that is broken up, the average of at least one of the neutrons thus produced finds another U-235, the reaction will go on all by itself. This is called a *self-sustained chain reacting pile*. Such reaction may go rather fast and produce much energy, although the effect is almost negligible compared with what happens in a piece of pure U-235, the atom bomb.

It is expected that further development of the chain reacting uranium pile will lead to the construction of a source of energy equivalent to that of huge hydroelectric plants. The wartime use of the uranium pile is, however, quite different. And this leads us to part three of the problem.

In a chain reacting pile a large fraction of the neutrons do not collide with U-235 but are captured by the more abundant U-238. Thus they are "lost" as far as the operation of the pile is concerned. It was discovered, however, that this U-238, after it captures a neutron, changes slowly over into a new substance, now called plutonium. But this plutonium has the same fission properties as U-235 and can thus be used in a bomb. Moreover, it is very much simpler to separate the plutonium from uranium than to separate U-235 from U-238.

The pile can still be used for another purpose. Most substances become radioactive when their atomic nuclei capture a neutron. Radioactive substance can be used for research, for therapeutic treatment, but in too large doses are dangerous because of their destructive radiations. In the chain reacting pile there is a huge number of neutrons produced and substances placed near or in the pile can be made strongly radioactive. In addition, the parts into which the nucleus of U-235 splits in the chain reaction are also strongly radioactive elements. Thus a uranium pile produces highly intense radioactive materials, equivalent to tons of radium, which in wartime could be used as poisons.

Recapitulating, we have the following three phases of the uranium problem:

(1) An atomic bomb made of pure U-235. This rare substance is extremely difficult to separate from U-238.

(2) A chain reacting pile made of ordinary uranium, a mixture of 0.7% U-235 and the rest U-238, embedded in a special arrangement in "heavy water" or carbon. Such a pile can produce plutonium and also radioactive materials of high intensity. After some further development it will also be used as a source of industrial energy.

(3) A bomb made of plutonium. It is generally accepted now that it is much simpler to construct a chain reacting pile and produce and remove the plutonium than to remove U-235 from natural uranium.

Turning again to the essential terms with which the reader should be familiar, we have the following:

FISSION. Otto Hahn, a German chemist, discovered in December, 1938, that in certain experiments it is possible to break up the atomic nucleus of U-235. In this so-called "fission," an amount of energy is produced which is 100 million times larger than in ordinary burning. So far, only three substances are found to possess this property, U-235, plutonium, and U-233. The latter two can be made in quantity only in a uranium pile; they do not occur in appreciable amounts in nature, if at all.

251

A NEUTRON is one of the particles which form the building material of an atomic nucleus. This particle has about the same weight as the nucleus of a hydrogen atom (called proton), but it carries no electrical charge. The atomic nucleus of the rare "heavy hydrogen" is built up of a neutron and a proton. The nucleus of U-238 contains 92 protons and 146 neutrons, 238 particles in all. For U-235 it is 92 protons and 143 neutrons.

HEAVY WATER. This is a variety of ordinary water, chemically not distinguishable but about 10% heavier, because it contains hydrogen atoms of double the usual atomic weight. Heavy water occurs in minute quantities in ordinary water from which it is difficult to separate. For a long time a Norwegian hydroelectric plant was the only place where it could be made in quantities.

URANIUM PILE. This is an arrangement of pieces of uranium metal embedded in carbon or in heavy water. The total amount weighs many tons.

If properly designed a self-sustained reaction takes place in a pile which produces large amounts of energy. When the reaction gets out of control, the pile will explode with only moderate speed; it is definitely *not* a bomb. During the early stages of A-bomb research it was erroneously believed that a bomb would result from letting a pile reaction get out of control. The Germans maintained this wrong idea to the very end.

A uranium pile produces large amounts of artificial radioactive substances, equivalent to tons of radium. In wartime these might be used to poison enemy supplies or in the form of poison gas. The radioactive radiations have a deadly effect on living beings. In peacetime these radioactive substances can be used for medical treatments, like radium or X rays, and for important research.

The pile also produces plutonium, a substance which can be used in an A-bomb.

U-235. Ordinary uranium metal contains 0.7% of U-235. Uranium 235 is fissionable and if brought together quickly in the right amount it will explode with tremendous power and also emit deadly radiations, as well as produce artificially radioactive substances. It is extremely difficult to separate U-235 from the ordinary uranium metal.

PLUTONIUM. This is a substance with the same explosive properties as U-235. It does not occur in nature, but is produced in a uranium pile.

A CYCLOTRON is a machine for performing laboratory experiments on the nuclei of atoms. It was invented by Ernest O. Lawrence of the University of California. It consists primarily of a huge magnet and an ingeniously designed chamber in which light atomic nuclei can be given a very high speed. Several substances can be made radio-active by bombarding them in the beam of fast-moving atomic particles in a cyclotron. The amounts produced are, however, negligibly small compared with the amounts produced in a uranium pile. A cyclotron is an important piece of scientific laboratory apparatus. There were about twenty in operation in the United States, but none in Germany before the war; German physicists made the one in Paris work and also built one of their own in Heidelberg.

Index of Principal Names

(fp, Frontispiece)

(f, Foreword)

255

BIOGRAPHICAL NOTE

R.V. Jones, Emeritus Professor of Natural Philosophy at Aberdeen University, was involved in British Scientific Intelligence at the highest levels during the second world war: he was Scientific Officer in the Air Ministry, Assistant, and later Director, of Intelligence in the Air Staff. The editors have asked him to contribute an introduction to this new edition of *Alsos* because he was one of the two British officers in charge of the captured German nuclear physicists who were secretly interned at Farm Hall, an English country house in Godmanchester near Cambridge, England. Well-known to British television and radio audiences, Professor Jones was educated at Oxford and holds a Doctor of Philosophy degree in Physics. He is a Fellow of the Royal Society and has been the recipient of numerous awards and honors for his research on the development of scientific instruments.